Structural Dynamics in Earthquake and Blast Resistant Design

Structural Dynamics in Earthquake and Blast Resistant Design

B.K. Raghu Prasad
Professor (Retd), Department of
Civil Engineering, Indian Institute of Science
Bengaluru, India

CRC Press is an imprint of the
Taylor & Francis Group, an **informa** business

First edition published 2020
by CRC Press
6000 Broken Sound Parkway NW, Suite 300, Boca Raton, FL 33487-2742

and by CRC Press
2 Park Square, Milton Park, Abingdon, Oxon, OX14 4RN

© 2021 Taylor & Francis Group, LLC

CRC Press is an imprint of Taylor & Francis Group, LLC

Reasonable efforts have been made to publish reliable data and information, but the author and publisher cannot assume responsibility for the validity of all materials or the consequences of their use. The authors and publishers have attempted to trace the copyright holders of all material reproduced in this publication and apologize to copyright holders if permission to publish in this form has not been obtained. If any copyright material has not been acknowledged please write and let us know so we may rectify in any future reprint.

Except as permitted under U.S. Copyright Law, no part of this book may be reprinted, reproduced, transmitted, or utilized in any form by any electronic, mechanical, or other means, now known or hereafter invented, including photocopying, microfilming, and recording, or in any information storage or retrieval system, without written permission from the publishers.

For permission to photocopy or use material electronically from this work, access www.copyright.com or contact the Copyright Clearance Center, Inc. (CCC), 222 Rosewood Drive, Danvers, MA 01923, 978-750-8400. For works that are not available on CCC please contact mpkbookspermissions@tandf.co.uk

Trademark notice: Product or corporate names may be trademarks or registered trademarks, and are used only for identification and explanation without intent to infringe.

ISBN: 978-0-8153-7018-5 (hbk)
ISBN: 978-1-351-25052-8 (ebk)

Typeset in Nimbus
by Nova Techset Private Limited, Bengaluru & Chennai, India

Visit the Taylor & Francis Web site at
http://www.taylorandfrancis.com

and the CRC Press Web site at
http://www.crcpress.com

Dedicated to my dear divine gurus and my dear parents

Contents

Preface xi

Acknowledgments xiii

Author xvii

Symbols xix

Chapter 1 Introduction 1
    1.1 Types of Analysis 2
    1.2 Modelling of a Dynamic System 3
        1.2.1 Degrees of Freedom 4
    1.3 D'Alembert's Principle 4

Chapter 2 Single Degree of Freedom Systems 7
    2.1 Introduction 7
    2.2 Free and Forced Vibrations 7
        2.2.1 Free Vibrations 7
        2.2.2 Damped Free Vibrations 11
        2.2.3 Logarithmic Decrement 16
    2.3 Forced Vibration of a Damped Single Degree of Freedom 17
    2.4 A Single Degree of Freedom System Subjected to Support Motion 21
    2.5 Rayleigh's Method to Obtain Natural Frequency 23
    2.6 Response in Frequency Domain and Laplace Transformation 25
    2.7 Solved Problems 30
    2.8 Exercise Problems 40

Chapter 3 Two Degree of Freedom Systems 43
    3.1 Forced Response of Damped Two-Storey Buildings 46
    3.2 Exercise Problems 48

Chapter 4 Force Transmitted to the Support 49
    4.1 Exercise Problems 52

| Chapter 5 | Duhamel's Integral | 53 |

| Chapter 6 | Modal Analysis | 61 |

6.1 Multi-Degree of Freedom Systems Subjected to External Dynamic Forces: Modal Analysis 61
6.2 A Multi-Storey Building Subjected to Ground Motions: Modal Analysis 65
6.3 Solved Problems 68
6.4 Exercise Problems for Chapters 5 and 6 81

| Chapter 7 | Earthquake Resistant Design | 83 |

7.1 Introduction 83
7.2 Structural Analysis 83
7.3 Structural Model 84
7.4 Shear Building 84
7.5 Response Spectrum 85
7.6 Capacity Spectrum 98

| Chapter 8 | Inelastic Vibration Absorber Subjected to Earthquake Ground Motion | 107 |

8.1 Introduction 107
8.2 The Linear Elastic Vibration Absorber 108
8.3 The Hysteric Vibration Absorber 111
8.4 Structural Model and the Equations of Motion 112
8.5 Numerical Studies 113
8.6 Analysis of Results 114
    8.6.1 Response of the Absorber Mass 119
    8.6.2 Response History Curves 120
    8.6.3 Hysteric Energy Dissipation 124
    8.6.4 Influence of Viscous Damping 130
    8.6.5 Maximum Ductility Response Spectra 130
8.7 Conclusions 132

| Chapter 9 | Inelastic Torsional Response of a Single-Storey Framed Structure: Two Degree of Freedom System | 133 |

9.1 Introduction 133
9.2 Earthquake Response of Elastic Structure with Coupled Translational and Torsional Motions 134
9.3 Structural Model 136
9.4 Equations of Motion 139
    9.4.1 Solution of the Equations of Motion 142
    9.4.2 Parameters Considered in the Study 142
    9.4.3 Details of the Computer Program 144

Contents

		9.5	Discussion of Results..144
			9.5.1 Influence of Eccentricity Envelopes of Maximum Frame Ductility 144
			9.5.2 Influence of Yield Strength 146
			9.5.3 Influence of P-Δ Effect...........................149
			9.5.4 Influence of Strengthening the Exterior Frames....149
			9.5.5 Response History Curves........................151
			9.5.6 Energy Dissipation due to Hysteresis 156
			9.5.7 Maximum Ductility Response Spectra 161
		9.6	Summary and Conclusions 162

Chapter 10   Inelastic Torsional Response of a Single-Storey Framed Structure: Three Degree of Freedom System 165

     10.1 Introduction...165
     10.2 Structural Model 166
         10.2.1 Yielding Behavior 167
     10.3 Equations of Motion 170
     10.4 Solutions of the Equations of Motion.......................172
         10.4.1 Parameters Considered in the Study 173
     10.5 Discussion of Results...174
         10.5.1 Influence of Eccentricity.........................174
         10.5.2 Influence of Yield Strength $q_{iuo} = q_{ivo} = q_i$ and Period ($T_{iu} = T_{iv} = T_i$) 180
         10.5.3 Time-Response Curves 181
     10.6 Summary and Conclusions 183

Chapter 11   Earthquake Resistant Design as per IS 1893:2016................189

     11.1 Introduction...189
     11.2 Project - 01 189
         11.2.1 Introduction...189
         11.2.2 Floating Columns...................................191
         11.2.3 Soft Storey 197
         11.2.4 Building Asymmetric in Plan 197
         11.2.5 Mass Participation Factor.......................206
         11.2.6 Conclusions...208
     11.3 Project - 02 209
         11.3.1 Introduction...209
         11.3.2 Analysis...209
         11.3.3 Conclusions...217
     11.4 Problems 217

Chapter 12 Miscellaneous Aspects 225
    12.1 Introduction 225
    12.2 Retrofitting Methods in Reinforced Concrete Structures ... 225
        12.2.1 Structure-Level Retrofit 225
        12.2.2 Addition of Shear Walls 226
        12.2.3 Base Isolators 226
        12.2.4 Addition of Steel Bracing 226
        12.2.5 Member-Level Retrofit 226
    12.3 Response Spectrum Analysis Using Python 236
        12.3.1 About the Programming Language: Python 236
        12.3.2 History of Python 236
        12.3.3 Application of Python in Civil Engineering 237
        12.3.4 Python Architecture 237
        12.3.5 Python Libraries 239
        12.3.6 Static Loading Problem Using Python 240
        12.3.7 Free Vibration Problems 244
        12.3.8 Response Spectrum Analysis of Building Using Python 245
    12.4 Hybrid Building under Seismic Forces 274
        12.4.1 Introduction 274
        12.4.2 Types of Connections 274
        12.4.3 Earthquake Responses of Hybrid Building 275
    12.5 Analysis and Design of Blast Resisting Structures 277
        12.5.1 General Characteristics of Blast and Consequences on Structures 277
        12.5.2 Loading Effects Due to Blasts 278
        12.5.3 Criteria for Blast Resistant Design of Structures for Explosions above Ground 279
    12.6 Response of RCC Asymmetric Buildings Subjected to Earthquake Ground Motions 281
        12.6.1 Structural Modelling 282
        12.6.2 Modelling and Analysis of Structural Irregularities 283

References and Suggestions for Further Reading 297

Index 305

Preface

I postponed writing a book like *Structural Dynamics in Earthquake and Blast Resistant Design* because I want this book to be slightly different from the books which are already out there on the subject. After much thinking I found that not many in the profession really appreciate the relevance of structural dynamics in earthquake and blast resistant design. Therefore, I felt that I should write a book which contains rudimentary principles of structural dynamics relevant enough in earthquake and blast resistant design.

There is nothing new in this book other than known things rearranged. I have said this after remembering a similar statement by a genius no less than Blaise Pascal. I always liked to see structural dynamics in our daily life. I was fond of giving such examples in my graduate teaching at the Indian Institute of Science (IISc), Bangalore. I have cited such examples in the book.

I wish that graduate students, students beginning their research and, finally, consulting engineers who design structures to resist earthquake and blast will find this book useful. Of course, let me admit, it is a book on structural dynamics with a flavor of earthquake and blast resistant design, or it is a book on earthquake and blast resistant design with a flavor of structural dynamics.

Whenever I have come across many practicing construction engineers and, consulting structural design engineers, they have always told me that they require a simplified version of structural dynamics that explain the physics behind it relating to earthquake and blast resistant design. Both senior engineers as well as those starting their careers as structural design engineers do not find enough time to dwell on theory. They are in a hurry to complete designs and often depend blindly on commercially available software codes. Most of them may not appreciate the ramifications of free vibration data, modal vectors and so on. In view of the above, I felt like writing a book explaining only the essentials of structural dynamics related to earthquake and blast resistant design giving examples from practical structural design. At the same time, there is not much emphasis on simple design and detailing because the information is plenty available elsewhere. Therefore, this book, I would say, would bridge the gap.

Nevertheless, there is a detailed exposition of inelastic torsion of buildings asymmetric in plan, capacity spectrum, structural rehabilitation, hybrid buildings and some new software called Python. Also there is an example on blast resistant design, particularly of structures below ground level. For the latter the I have developed a new technique of imagining a fictitious vertical surface just close to the source of explosion which will facilitate computation of overpressures on the building under consideration.

Acknowledgments

The most difficult part of any book is acknowledging persons who have helped directly or indirectly while writing a book because there will be so many visible and also invisible.

This work takes inspiration from *Introduction to Structural Dynamics* by J.M. Biggs, a book I read even before I was formally taught. The book starts with a simple explanation of the physics behind structural dynamics and ends with the then future topic, the Matrix approach. I must concede my debt to the late Prof. K.T.S. Iyengar, Department of Civil Engineering, IISc, whose name was and is even now synonymous with structural mechanics and concrete. He inspired me to work for my PhD in the area of structural dynamics under the guidance of Prof. K.S. Jagadish, who later became one of my closest colleagues. He taught me structural dynamics, matrix analysis of structures and much more. He made the tough subject look so easy by highlighting the physics. My knowledge of inelastic vibration absorber and inelastic torsional response is due to him. Prof. R.N. Iyengar, another teacher-turned-colleague and currently Distinguished Professor, Centre for Disaster Mitigation, Jain University, rigorously taught me stochastic process, random vibrations and probabilistic methods of design. There were other great stalwarts in the department who inspired me. They were Prof. K. Chandrashekhara, the late Prof. S. Anantharamu and the late Prof. Prakash Desayi. Prof. A. Sridharan of the Department of Geotechnical Engineering, former Deputy Director of IISc, even today encourages me very affectionately to write papers and textbooks, and Prof. K.S. Subba Rao of the same department is also a great moral support for me. My sincere thanks go to all of them.

My students, innumerable in number at IISc, the Oxford College of Engineering (TOCE) and K.S. School of Engineering and Management, Bangalore (KSSEM), have always appreciated my teaching and guidance and thus always have encouraged me. Particularly, Mr. Shreyas Srivatsa, currently a doctoral student in Krakow, Poland, encouraged and pushed me to complete the book. There were some setbacks while writing, and he helped me in preparing the typed script. He typed in part and later as he had to go to Poland he, with great difficulty, requested his friend Yoganarasimha B.G., Administrative Assistant, Indian Academy of Sciences, Bengaluru, to continue typing. Mr. Yoganarasimha, who, in a short time, had to learn LaTeX code, which was a requirement needed to complete the book, has to be specially appreciated. Even from Poland, Mr. Shreyas would inquire about the progress of the book and help further by working out a number of examples and also proofread. Mr. Karan Davda, my current graduate student in the Department of Civil Engineering at TOCE, Bangalore, took care of typesetting the manuscript. He also worked out solutions to some problems. Without the timely help of the above three dedicated persons, the book could not have been completed. Mr. Karan also helped me in transferring relevant parts of our collaborative work on capacity spectrum for inelastic torsional response from his dissertation. He also proofread the manuscript along with me. My thanks to them. I would also like to thank the other current students of the Civil Department at TOCE, namely Ms. Sharlin Sheeba, Mr. Amir Khan, Ms. Swati Ricke, and Mrs. Gayathri R., who similarly agreed to share results from collaborative work toward their dissertations, which were produced by the students themselves under the guidance of myself and Prof. Amarnath K., Director, Oxford Institutions. My thanks also go to Dr. Malleshiah and Mr. Shivanand of TOCE, Bangalore. My thanks are to Dr. Sriman Narayan H.N., Principal consultant, Indigo Consultants who typed and proof read a part of the book and Sri Anil Hosur, Senior Engineer, M/s Sundaram Architects who gave hints on file management on laptop.

Mr. Digvijay Patankar, my former student at IISc and currently a PhD student at IIT Roorkee, lent a portion of his notes and also encouraged me. My thanks are due to him.

Prof. Akella Vijayalakshmi, Head of the Department of Civil Engineering, KSSEM, has provided solutions to several problems. My thanks go to her. Thanks are also to Prof. Veerendrakumar and Prof. Arekal Vijay of KSSEM.

Mr. H.N. Renuka Prasad, founder and CEO, Prasad Consultants, consults me and seeks my advice for structural design involved in his projects. I have borrowed details of those projects. My thanks go to him.

Mr. Vijayaraj, N.K., practicing structural engineer, also helped me in proofreading. My thanks go to him.

Now, more importantly, I have to sincerely thank my wife, Dr. Shailaja Raghu Prasad, who constantly encouraged me and also suffered through several moments of my preoccupation with the book. She, being a teacher herself, kept me disciplined with a strict regime without which I might not have completed the work. She also proofread the manuscript. Her parents, particularly her father the late Sri B.K. SubbaRao, himself a civil engineer, would often suggest that I write a book. My thanks are to them. In addition, I thank my uncles and aunts, my sister Dr. Shashi Prabha Dasgupta, and Prof. Dipak Dasgupta. The prayers of my old friend Sri SriHari for a successful completion of the book were necessary, and my grateful thanks are to him.

I have to sincerely acknowledge the constant encouragement of my two sons, Dr. Radhakrishna Bettadapura, Senior Director, Strand Life Sciences, Bangalore, along with his wife Ms. Sucharita, and Mr. Shivatejas, Senior Lead Engineer, Qualcomm, Bangalore, as well as his wife Ms. Sumedha.

Dr. Radhakrishna, who is himself an avid reader of science and a writer, would very often inquire about the book's progress, and Mr. Shivatejas, who is familiar with 5G speed stochastic signals, would push me. He, along with Ms. Sumedha, also suggested certain corrections for the Preface and the Acknowledgments.

Acknowledgments

On a personal note, I believe in divine help for any human endeavor. It is beyond science and can only be experienced.

I have to mention exclusively the divine force which the human eye cannot perceive. Particularly, an engineer, who is to be responsible for safety of life and the environment, who has to deal with a vast number of uncertainties, assumptions, experimental errors and noises, has to finally pray to the divine force. Therefore, last but certainly never least, I mention exclusively my divine Gurus, Shirdi Sri Sai Baba and H.H. Sri Saipandananda Radhakrishna Swamiji, and my beloved parents, namely my father, the late Sri B.N. Krishna Swamy, advocate and public prosecutor who inspired and blessed me to do my PhD at IISc, and my mother the late Smt. Kamala Krishna Swamy, a science and mathematics graduate and teacher, who was very proud of my tenure at IISc, and who inspired and blessed me to write the book. Sadly, both of them did not live to see the completion of it. No words are adequate to express my fond gratitude to them. Therefore, I have dedicated the book to them.

I am aware that I have written a very lengthy list of names to be thanked. It is by no chance of my presumption of any monumental nature of the book. Even then, whatever little good the reader finds is due to them and whatever is uninteresting is due to me.

I thank Gagandeep Singh, the Development Editor, Michele Dimont, the Project Editor and their team, at CRC Press as well as Kavitha, Project Manager and Ganesh at Nova Techset, India for their meticulous, committed and excellent work but for which the book would not have seen the light of the day.

Author

Professor B.K. Raghu Prasad was awarded a PhD from the Indian Institute of Science in 1979. His thesis was titled, "Some Problems in the Inelastic Response of Structures Subjected to Earthquakes." He worked on designing a top storey of any structure to undergo yielding such that it absorbs the energy of an earthquake to the extent that it can keep the bottom storey(s) relatively within elastic limits. The top storey could also be the topmost storey of any tall building. The storey that absorbs the energy has to be necessarily a bilinear hysteretic yielding structure. It could be deliberately designed that way so that it could be called an expendable weak storey which could be easily rebuilt after the earthquake while retrofitting the structure. In the process, the bottom storey(s) need not be disturbed and could be retained. Similarly, he worked on structures asymmetric in plan which undergo torsion along with translations along horizontal directions.

Later, Dr. Prasad had a brief stint in practical designs. He was a senior executive at Canadian METCHEM consultants working on a Kudremukh iron ore project in India and later with Engineers India Ltd., New Delhi. He came back to the Department of Civil Engineering, Indian Institute of Science, Bangalore as a faculty member in 1982–2011 untill he retired as a professor in the department. During the course of his stay, he was also the chairman of the department. He was also a chairman of the Centre for Campus Management and Development (CCMD) looking after the designs and maintenance of the structures on the campus. He was also the Secretary and Vice Chairman of GATE examinations. Finally he served as Advisor, CCMD after retirement till 2016. His early education was at the national high school and college, BMS College of Engineering and University, Visvesvaraya College of Engineering, all in Bangalore.

He taught various subjects such as Structural Dynamics, Structural Design for Dynamic Loads, Matrix Analysis of Structures, Design of Shells, Design of Concrete and Prestressed Concrete Structures and Fracture Mechanics of Concrete. He has guided 27 PhD, 8 MSc and over 30 ME students, including 4 students from Norwegian Institute of Science and Technology, Trondheim, Norway, for their respective degrees. He has published over 250 papers in peer-reviewed journals and international and national conferences. He has travelled all over the world and delivered talks at premier institutes such as MIT, USA, UMIST, UK, and Norwegian Institute of Science and Technology, Trondheim, Norway. He was a consultant for several nationally important projects taken on by the Institute for Government, both central and state, including nuclear projects from BARC and power projects as well as a project for the Indian Navy. He visited Japan as an INSA-JSPS Fellow in 1992 and 1993. Finally, he was advisor at the Institute for CCMD after retirement until 2016. Even now, he is still active as a Visiting Professor at two private engineering colleges, the Oxford

Engineering College, Bangalore and K.S. School of Engineering and Management, Bangalore. There also he has been guiding students on their ME dissertations and PhD degrees. So far, in those institutions, he has guided about 30 ME students and 2 PhD students in obtaining their degrees. Further, his advice is sought on structural designs of tall buildings particularly for seismic considerations by several agencies and consultants such as Sundaram Consultants and HNR Prasad Consultants, Bangalore.

Symbols

Static and Dynamic Analysis

DOF - Degrees of freedom

F(t) - External force

m - Mass

k - Stiffness

c - Damping

$\ddot{x}$ - Acceleration

$\dot{x}$ - Velocity

x - Displacement

SDF - Single degree of freedom system

ω - Natural frequency, radians per sec

T - Time period

f - Natural frequency in cycles per sec or Hz

A - Amplitude of oscillation

θ - Phase angle

c_{cr} - Critical damping constant

ω_d - Damped natural frequency

η - Damping ratio

δ - Logarithmic decrement

Ω - Forcing frequency

r - Frequency ratio

x_{st} - Static deflection

F(t) dt - Impulse force applied on masses over a small duration t

t_a - Rise time

t_d - Time duration

$\bar{m}_1, \bar{m}_2$ - Modal masses

$\bar{k}_1, \bar{k}_2$ - Modal stiffnesses

$\bar{F}_1, \bar{F}_2$ - Modal forces

SRSS - Square root sum of squares

RMS - Root mean square

$x_s(t)$ - Base motion

x_r - Relative displacement

P_{fn} - Modal participation factor in n^{th} mode

[k] - Stiffness matrix

[m] - Mass matrix

$\dot{x}_{pv}$ - Max. pseudo-velocity

$\ddot{x}_{pa}$ - Max. pseudo-acceleration

x_{rm} - Maximum deformation

x_{gm} - Peak ground displacement

$\ddot{x}_{gm}$ - Peak ground acceleration

$\dot{x}_{gm}$ - Peak ground velocity

x_g - Ground displacement

m_n - n^{th} modal mass

k_n - n^{th} modal stiffness

x_n - n^{th} modal displacement

F_n - n^{th} modal force

$[\dot{x}]$ - Velocity vector

[x] - Displacement vector

[F(t)] - Force vector

S_d - Spectral displacement

S_a - Spectral acceleration

V_B - Base shear

μ - Ductility factor

R_μ - Reduction factor due to ductility

T_c - Characteristic period of ground motion

I_o - Immediate occupancy

LS - Life safety

CP - Collapse prevention

SBC - Safe bearing capacity of soil

CFST - Concrete filled steel section

Symbols xxi

p_s - Overpressure

p_{s_o} - Peak initial overpressure

q_o - Maximum dynamic pressure

P_d - Dynamic pressure

ρ - Density of air

V - Velocity

C_d - Drag coefficient

e_x - Eccentricity in X-direction

e_y - Eccentricity in Y-direction

MF - Magnification factor

DLF - Dynamic load factor

Q - Quality factor

g - Acceleration due to gravity

Δ - Deflection

$\ddot{x}_s$ - Support acceleration

E - Total energy

K.E. - Kinetic energy

P.E. - Potential energy

x_m - Max. displacement

s - Laplace variable

Lx(t) - Laplace transformation

$Z(i\Omega)$ - Impedance function

G(s) - Transfer function

2DOF - Two degrees of freedom system

$\frac{\bar{x}_1}{\bar{x}_2}$ - 1st mode eigenvector

$\bar{x}_s$ - Amplitude of harmonic support displacement

T_r - Transmissibility

Chapter 8

z_1 - Non-dimensional displacement of the bottom storey relative to the base

z_2 - Non-dimensional displacement of the top storey relative to the bottom storey

Y_I, Y_{II} - Relative yield displacements of the bottom and top storeys, respectively

$P(z_1)$, $P(z_2)$ - Non-dimensional resistance forces of the bottom and top storeys, respectively

$S_1 - c_1 = 2\sqrt{K_1 m_1}$

$S_2 - c_2 = 2\sqrt{K_2 m_2}$

c_1, c_2 - Damping constants of the bottom and top storeys, respectively

$\tau = \omega_1 t$ - Damping constants of the bottom and top storeys, respectively

$\omega_1 = \sqrt{K_1 m_1}$

$\omega_2 = \sqrt{K_2 m_2}$

$F = \omega_2/\omega_1$ = Ratio of frequencies

$Y = Y_{II}/Y_I$ = Ratio of yield displacements

$\mu = m_2/m_1$ = Ratio of masses

T - Linear elastic period of the bottom storey

q_Y - Acceleration required to cause yielding of the bottom storey expressed as a fraction of the acceleration due to gravity, g

g - Acceleration due to gravity

D_1 - Ductility factor of the bottom storey

D_2 - Ductility factor of the top storey

$\overline{D}_o$ - Maximum ductility of the structure with the absorber removed (S.D.F.)

$\overline{D}_1, \overline{D}_2$ - Maximum ductilities of the bottom and top storeys, respectively

T_B - Dominant period of vibration of the bottom

T_T - Dominant period of vibration of the top storey

H_B - Total hysteretic energy absorbed by the bottom storey up to the end of the response

H_T - Total hysteretic energy absorbed by the top storey up to the end of the response

E_B - Elastic strain energy that can be stored in the bottom storey

Chapter 9

a - Span of the portal frames

b - Length of the roof slab (dimension along the Y-direction)

h - Rise of the portal frames

G - Mass center

S_C - Center of stiffness

e - Eccentricity (expressed as a ratio, e/b)

Symbols

N - Number of portal frames

M - Total mass of the entire structure

I – Polar moment of inertia of the roof slab about a vertical axis through the mass center

$r-g$ - Polar radius of gyration of the slab about a vertical axis through the mass center

m_i - Mass supported by the ith frame

r_{fi} - Distance of the ith frame from the mass center measured along the Y-axis

U - Translational displacement of the mass center along the direction of the X-axis relative to the ground displacement

Φ - Rotational displacement of the roof slab about a vertical axis through the mass center

R_{fi} - Resistance of the ith frame - a function of the relative displacement U_{fi}, relative velocity, U_{fi} and time, t

C_{fi} - Damping constant of the ith frame

$U_{fi} - U + r_{fi}\Phi$ - Displacement of the ith frame along the direction of the X-axis

U_s - Ground of support displacement along the X-axis

Z - Normalized translational displacement of the mass center along the X-axis

R_{fin} - Normalized resistance of the ith frame

S_{fir} - Ratio of stiffness of the ith frame to that of frame, 1

Y_{fir} - Ratio of yield displacement of the ith frame

η_{fir} - Ratio of mass carried by the ith frame to total mass of the complete structure

U_{fin} - Normalized displacement of the ith frame

C_{f1r} - Viscous damping ratio of frame, 1

D_{fir} - Ratio of damping constant of the ith frame to that of frame, 1

r_{fig} - Normalized distance of the ith frame from the mass center

r_{gn} - Normalized polar radius of gyration

U_{f1y} - Yield displacement of frame 2 along the X-direction

K_{f1} - Stiffness of frame, 1

C_{f1} - Damping constant of frame, 1

q_{f1y} - Yield acceleration of frame, 1 (expressed as a fraction of g)

q_{fiy} - Yield acceleration of frame, I (expressed as a fraction of g)

g - Acceleration due to gravity

$Q_{f1y} - \omega_{f1}^2 U_{f1y}$

$\omega_{f1} - \sqrt{K_{f1}/M}$

$\tau - \omega_{f1}t$ - non-dimensional time

f_{t1}, f_{t2} - linear frequencies of the complete structure in the I and II modes, respectively

T_{fi} - Linear elastic period of the ith frame

D_{fi} - Ductility factor of the ith frame

$\overline{D}_{fi}$ - Maximum ductility factor of the ith frame

N_{fi} - Number of yield level crossing of the ith frame

P - Vertical carried by each of the columns in the ith frame

Δ - Horizontal sway of the portal frames (same as the lateral displacement of the frame, U_{fin})

Hz - Hertz

H_{fi} - Total hysteretic energy dissipated upto the end of the quake duration in the ith frame

H_{fio} - Total hysteretic energy dissipated upto the end of the quake duration in the ith for the zero-eccentricity case

$H_{fc}(\tau)$ - Cumulative hysteretic energy dissipated in the ith frame upto time

E_L - Elastic strain energy which can be stored in the ith frame of the structure with $q_{fiy} = 0.14g$

Chapter 10

Dimension of the roof slab along the X (also Y) direction

G - Mass center

S_C - Center of stiffness

e_x, e_y - Eccentricities along the X- and Y-axes, respectively (it is assumed that $e_x = e_y = e$)

N_C - Number of columns

M - Total mass of the entire structure lumped at the floor level

I - Polar moment of inertia of the roof slab about a vertical axis through the mass center

r_g - Polar radius of gyration of the slab about a vertical axis through the mass center

K_{iu}, K_{iv} - Elastic stiffnesses of the ith column along the X- and Y-axes, respectively ($K_{iu} = K_{iy}$ because columns are symmetrical in cross-section)

Symbols

r_{iu}, r_{iv} - Distance of the ith column from the mass center along the X- and Y-axes, respectively

U, V - Translational displacements of the mass center along the directions of the X- and Y-axes, respectively, relative to the ground displacement

Φ - Rotational displacement of the slab about a vertical axis through the mass center

R_{iu}, R_{iv} - Resistance of the ith column along X- and Y-axes, respectively – function of relative displacements, relative velocities and time

$U_i = U + r_{iv}\Phi$ - Displacement of the ith column along the X-axis, relative to the ground displacement

$V_i = V - r_{iu}\Phi$ - Displacement of the ith column along the Y-axis, relative to the ground displacement

C_i - Damping constant of the ith column

U_s, V_s - Ground or support displacement along the X- and Y-axes, respectively

$U_n - U/U_{io}$ - Non-dimensional displacement of the mass center along the X-axis

$V_n - V/V_{io}$ - Non-dimensional displacement of the mass center along the Y-axis

$R_{iun} - R_{iu}/R_{iuo}$ - Normalized resistance of the ith column along the X-axis

$R_{ivn} - R_{iv}/R_{iyo}$ - Normalized resistance of the ith column along the Y-axis

$U_{in} - U_i/U_{io}$ - Normalized displacement of the ith column along the X-axis

$V_{in} - V_i/V_{io}$ - Normalized displacement of the ith column along the Y-axis

$S_{ic} = \frac{C_i}{2\sqrt{K_{iu}M}}$ - Non-dimensionalized damping ratio

$r_{gnu} = r_g/U_{io}$ - Polar radius of gyration of the slab about a vertical axis through the mass center, normalized with respect to yield displacement of the ith column along the X-axis

$r_{gnv} = r_g/V_{io}$ - Polar radius of gyration of the slab about a vertical axis through the mass center, normalized with respect to yield displacement of the ith column along the Y-axis

$r_{igu} = r_{iu}/r_g$ - Normalized distance of the ith column from the mass center measured along the X-axis

$r_{igv} = r_{iv}/r_g$ - Normalized distance of the ith column from the mass center measured along the Y-axis

q_{iuo}, q_{ivo} - Yield acceleration of the ith column along the X- and Y-axes, respectively (they are expressed as fractions of acceleration due to gravity, it is assumed that $q_{iuo} = q_{ivo} = q_i$)

g - Acceleration due to gravity

U_{io}, V_{io} - Yield displacement of the ith column along the X- and Y-axes, respectively, corresponding to uniaxial loading (it is assumed that $U_{io} = V_{io}$)

R_{iuo}, R_{ivo} - Yield displacement of the ith column along the X- and Y-axes, respectively, corresponding to uniaxial loading (it is assumed that $R_{iuo} = R_{ivo}$)

$\omega_n = \sqrt{K_{iu}/M}$

$\omega_v = \sqrt{K_{iv}/M}$

t = Time in seconds

$\tau = \omega_u t = \omega_v t$ - Non-dimensional time

T_{iu}, T_{iv} = Linear elastic periods of the ith column in seconds; along the X- and Y-axes, respectively (it is assumed that $T_{iu} = T_{iv} = T_i$)

D_{iu}, D_{iv} - Ductility factors of the ith column along the X- and Y-axes, respectively

$\overline{D}_{iu}, \overline{D}_{iv}$ - Maximum ductility factors of the ith column along the X- and Y-axes, respectively

$\overline{D}_{max}$ - Largest value of the maximum ductilities $\overline{D}_{iu}$ and $\overline{D}_{iv}$ of all the columns

$\overline{D}_{min}$ - Smallest values of the maximum ductilities $\overline{D}_{iu}$ and $\overline{D}_{iv}$ of all the columns

1 Introduction

In structural engineering practice, design for dynamic loads is important. Most common dynamic loads are those due to earthquake ground motion, wind, blast and machinery, etc. Vehicles travelling on a rough surface also undergo dynamic loading and in turn transfer those dynamic effects to the surface of the vehicle in which the passengers are seated. But the problem which generally an engineer encounters is the decision regarding whether the loads are dynamic or mere static.

As it strikes to common sense, generally speed controls dynamic effects. It is true that in all our experiments we do control the rate of loading which is also the speed of loading. Speed is a relative variable and, therefore, with respect to what we compare the rate of loading, to judge it as fast or slow is an important question. Here, the concept of theory of vibrations is considered. A system parameter called natural period (T in seconds), which is the reciprocal of natural frequency (f, in Hertz), comes in handy as a yard stick to compare the rate of loading as sudden or gradual. Various types of dynamic loading, like suddenly applied load, ramp load, pulse loads of different shapes, sinusoidal and random loadings, are common in practice.

A machine foundation supporting rotating machinery having an eccentric mass on the rotor will cause a sinusoidal force on the foundation. A tall building or a tall tower subjected to along and across wind oscillations is an example of sinusoidal load due to wind. The wind at certain velocities will cause resonant condition of the tower. Earthquake ground motion is an example of random base motion, which can cause severe damage to a structure if not properly designed. A distant blast or an explosion causes a triangular pulse load on a structure which is in the vicinity thus intercepting the over-pressure waves from the blast. Sudden fall of an object could be an example of an impulse. A suddenly applied load held constant over time can be a step load. Similarly, one can find examples of different types of pulse loadings. In all the examples cited above, the first mode natural period, T becomes important. Either the duration of the pulse in the case of pulse loads or the period of oscillation of the force in the case of sinusoidal load or reciprocal of the number of zero crossings per second in the case of random loading is compared with the natural fundamental period of the system. Their ratio is an important parameter which decides the severity of vibration.

1.1 TYPES OF ANALYSIS

Static Analysis and Dynamic Analysis

In any problem dealing with either statics or dynamics, there is always a system with an input and output. Why we need to know static analysis while understanding dynamics is because in many cases a dynamic problem is solved as an equivalent static problem by converting dynamic loads as equivalent static loads and performing static analysis (Figs. 1.1 and 1.2).

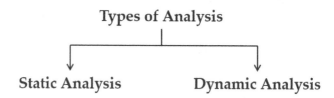

Generally, the system and input are known to obtain the required output. Input consists of geometrical as well as material properties, boundary conditions and forces acting on the system (Fig. 1.1). The outputs generally are bending moments, shear and axial forces, displacements, stresses and strains. The system is defined by appropriate system equations which could either be differential or integral equations. The differential equations will be ordinary in the case of discrete systems and partial in the case of continuous systems. In the case of stochastic systems, either the input or the system or both can be stochastic which obviously means that the output will always be stochastic. In dynamic analysis, the input forces are time varying. In linear systems, system properties remain constant and the responses have linear variation. In non-linear systems, the response values will be non-linear. Further, there could be non-linear elastic as well as inelastic systems. In non-linear elastic systems, the response is non-linear from the beginning but can come back to the original position on the same path when the forces are released; while in the non-linear inelastic systems, the system is non-linear right from the beginning and the system does not come to original position along the same path when the loads are released, thus expending some energy in the process. There are non-linear hardening and softening systems. Yet another important fact to note is that even a dynamic problem can be treated as an equivalent static problem by the use of dynamic load factors.

Some examples of problems in structural dynamics are listed below:

1. Effects of earthquake ground motions on structures.
2. Wind effect.
3. Vehicle moving on a bridge causing vibrations of the bridge.
4. Vehicle negotiating a road hump suffering vibrations.
5. Foundation of machine.
6. A passenger train car undergoing vibration causing discomfort to the passengers due to roughness of the rail.

Introduction

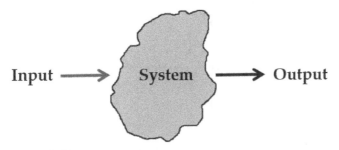

FIGURE 1.1: System subjected to input and providing an output

1.2 MODELLING OF A DYNAMIC SYSTEM

The dynamic response of a system can be defined by three quantities viz. displacement, velocity and acceleration. Knowing any one of the three, the remaining quantities can be obtained by either differentiation or integration depending on the case (Fig. 1.2). If displacement is determined first, the velocity and acceleration can be obtained by successive differentiation with respect to time. On the other hand, if acceleration is determined first, velocity and displacement can be obtained by successive integration with respect to time once and twice, respectively. In order to obtain the above quantities, the position of the system at any instant of time with respect to a reference coordinate system or in other words, displacement of the mass from a fixed

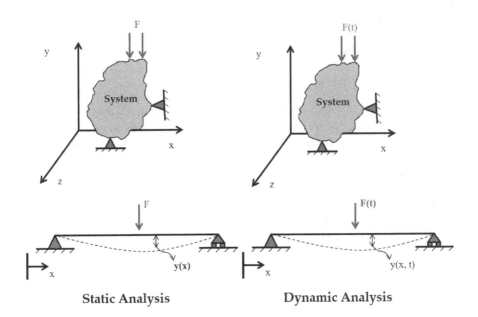

FIGURE 1.2: Static and dynamic cases

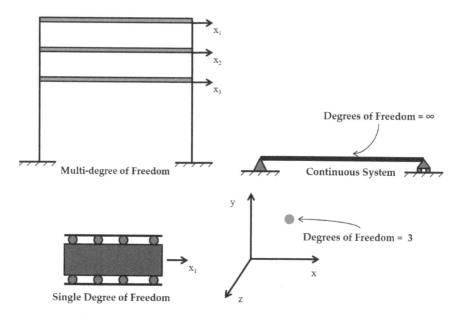

FIGURE 1.3: Examples of discrete and continuous systems and their corresponding degrees of freedom

reference XYZ has to be uniquely and independently defined obviously by X, Y and Z coordinates, respectively (Fig. 1.3). This can also be done in polar coordinates. The independent coordinate required to define the position of the vibrating mass is called generalised coordinate.

1.2.1 DEGREES OF FREEDOM

Degrees of freedom (DOF) are defined as the minimum number of independent generalized coordinates required to define the position of a body in motion. Degrees of freedom are finite in discrete systems and infinite in continuous systems (Fig.1.3).

1.3 D'ALEMBERT'S PRINCIPLE

The D'Alembert principle is that the force acting on a vibrating system, viz. reversed effective force which is inertia force considered to be acting in the opposite direction of motion, the spring and damping forces along with the external forces together, keep the body in a state of dynamic equilibrium (Biggs, J. M., 1964).

Using the above principle, the equation of motion can be written in a simpler form, as if the system is in static equilibrium, the term which is the inertial force is imagined to be acting in the negative x-direction opposite to the direction of acceleration as shown in Fig. 1.4. All the forces are algebraically summed up as in Equation 1.1.

Introduction

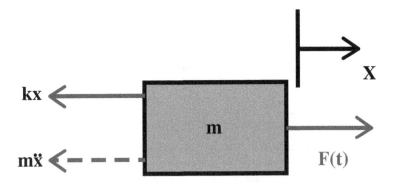

FIGURE 1.4: Free body diagram of the single degree of freedom system

The inertia force by D'Alembert's principle is shown in the direction opposite to the external force acting on the mass. kx is the spring force which will be obviously opposite to the direction of motion. Thus, writing as if the system is in static equilibrium we get,

$$F(t) - m\ddot{x} - kx = 0 \tag{1.1}$$
$$\therefore m\ddot{x} + kx = F(t) \tag{1.2}$$

If damping is considered, the additional opposing force is $c\dot{x}$.

Now as before summarizing up all the forces, where now, in addition to the opposing spring force -kx, we get obviously opposing damping force $-c\dot{x}$, thus we get

$$F(t) - m\ddot{x} - c\dot{x} - kx = 0$$

$$Thus, m\ddot{x} + c\dot{x} + kx = F(t)$$

The above way of writing the equation is convenient for those who are familiar with writing static equilibrium equation.

In analyzing structures for dynamic loads, modelling the structure either as a discrete system or as continuous system is necessary. In a discrete system, the masses are lumped at discrete points and the generalized coordinates are defined at the masses. Springs whose stiffnesses are obtained from the stiffness properties of the real structure are used to connect the masses as well as the boundary point. In fact, the popular finite element method divides a continuum to several elements and the corresponding masses are considered to be lumped at the mass center of the element. The generalized coordinates are defined at the respective mass center. Therefore the number of generalized coordinates considered for each of the masses multiplied by the number of elements will give the total number of degrees of freedom. When there is only one generalized coordinate required to define the system, it is called a single degree of freedom system. When there are more than one degree of freedom, it is called multi-degree of freedom system. Sometimes, simple individual members like beams and columns with various types of boundary conditions are solved considering distributed mass (continuous system) in which case one has to solve a partial differential equation of motion. In the case of lumped mass system, an ordinary differential equation has to be solved.

It may be appropriate to mention here although it may be repeated again under multi-degree of freedom systems that there will be as many equations of motions as the number of degrees of freedom. They will only be coupled either in stiffness or mass or in both. The coupling depends on the choice of the coordinates and it is not unique. If the masses are lumped, then the coupling will be only in stiffness and not in mass. If the mass of an element, say a beam element is expressed as consistent mass matrix, then the mass matrix will not be diagonal, but fully populated.

Till now, we have learnt how to write the equation of motion (can also be called differential equation of equilibrium) for one degree of freedom systems. It could be subjected to standard forces such as sinusoidal (harmonic) and step functions. There are more types of forces like pulse of different shapes and could be even an arbitrary shape which cannot be defined by any mathematical function and finally, random sinusoidal function like we observe in the earthquake ground acceleration. To obtain a solution of the forced part of the equation for such loads, an integral called Duhamel's integral is employed.

2 Single Degree of Freedom Systems

2.1 INTRODUCTION

2.2 FREE AND FORCED VIBRATIONS

Vibration during which no external force F(t) is applied on the system is called "free vibration." If an external force is operative during the vibration of the system, then it is called "forced vibration." A one storey building frame and the water tank shown in Fig. 2.1 can be modelled as a single degree of freedom (SDF), provided the mass of the building is lumped at the floor level and the rotations of the columns are assumed to be zero (restrained at floor level). The only degree of freedom here is x.

2.2.1 FREE VIBRATIONS

$$F(t) = 0 \quad Forcing\ term. \tag{2.1}$$

$$c = 0 \quad Undamped\ system. \tag{2.2}$$

Equation of motion (EOM)

$$m\ddot{x} + kx = 0 \tag{2.3}$$

The general solution could be of the exponential form (Leonard Meirovitch, 1986).

$$x = Ae^{st}, \tag{2.4}$$

Where A is any constant and s is the exponent.

Substituting the solution in the EOM (2.3)

$$Ams^2 e^{st} + Ake^{st} = 0 \tag{2.5}$$

$$ms^2 + k = 0 \tag{2.6}$$

The above equation is called the characteristic equation. The roots of which are called the eigenvalues from which the natural frequencies are obtained.

$$s_1 = i\sqrt{\frac{k}{m}} \tag{2.7}$$

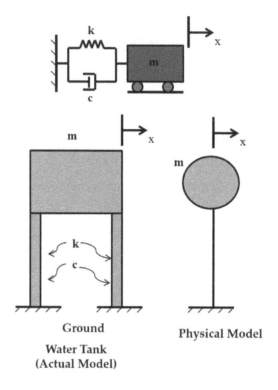

FIGURE 2.1: Modelling procedure illustration for simple systems with single degree of freedom

$$s_2 = -i\sqrt{\frac{k}{m}} \quad (2.8)$$

$$i = \sqrt{-1} \quad (2.9)$$

$$x = A_1 e^{s_1 t} + A_2 e^{s_2 t} \quad (2.10)$$

$$x = A_1 e^{it\sqrt{\frac{k}{m}}} + A_2 e^{-it\sqrt{\frac{k}{m}}} \quad (2.11)$$

$$Let, \sqrt{\frac{k}{m}} = \omega \quad (2.12)$$

Substituting,

$$x = A_1 e^{i\omega t} + A_2 e^{-i\omega t} \quad (2.13)$$

Single Degree of Freedom Systems

We know that,

$$e^{i\theta} = \cos\theta + i\sin\theta \tag{2.14}$$

$$e^{-i\theta} = \cos\theta - i\sin\theta \tag{2.15}$$

$$x = A_1(\cos\omega t + i\sin\omega t) + A_2(\cos\omega t - i\sin\omega t) \tag{2.16}$$

$$x = (A_1 + A_2)\cos\omega t + i(A_1 - A_2)\sin\omega t \tag{2.17}$$

$$C_1 = (A_1 + A_2) \tag{2.18}$$

$$C_2 = i(A_1 - A_2) \tag{2.19}$$

$$x = C_1\cos\omega t + C_2\sin\omega t \tag{2.20}$$

Now, we see that ωt takes the place of an argument of harmonic functions $\cos\omega t$ and $\sin\omega t$. Therefore, ω now can be defined as natural frequency of the oscillations $x(t)$. Therefore, natural frequency of the system can be defined as $\sqrt{\frac{k}{m}}$, k is the spring constant or stiffness and m is the mass of the system. The linear motion of a point lying on a vector rotating about a point at ω rad/sec can also be shown to be oscillatory of the simple harmonic type and therefore ω is also called circular frequency expressed as rad/sec. As 2π radians are equal to one cycle ω radians will make $\omega/2\pi$ cycles/sec denoted by f, cycles/sec or f, Hertz (Hz) named after famous scientist Hertz. T is the time required for one cycle or 2π radians as shown in Fig. 2.2.

$$\omega = \frac{2\pi}{T} \tag{2.21}$$

Also,

$$2\pi f = \omega \text{ and } \frac{1}{f} = T \tag{2.22}$$

Also, keeping $A_1 + A_2 = A\cos\theta$

$$i(A_1 - A_2) = A\sin\theta$$

Where A_1, A_2 and A are all constants of integration

$$x = A\cos(\omega t - \theta)$$

where, θ is called the phase angle and A is called the amplitude of oscillation.

To determine the constants of integration, we make use of the initial conditions at t = 0

$$x(0) = x_0 \tag{2.23}$$

$$\dot{x}(0) = \dot{x}_0 \tag{2.24}$$

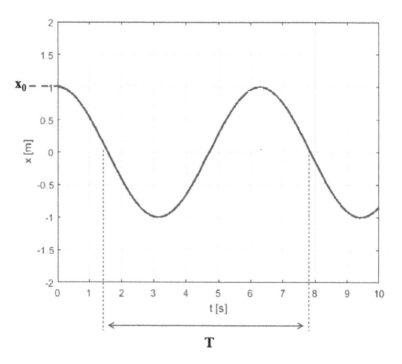

FIGURE 2.2: Displacement response of an undamped one degree of freedom system subjected to initial condition $x(0) = x_0$ and $\dot{x}(0) = \dot{x}_0 = 0$

Equation 2.20 now is,

$$x = x_0 \cos\omega t + \frac{\dot{x}_0}{\omega} \sin\omega t \qquad (2.25)$$

Let,

$$x_0 = A\cos\theta \qquad (2.26)$$

$$\frac{\dot{x}_0}{\omega} = A\sin\theta \qquad (2.27)$$

$$x = A\cos\theta\cos\omega t + A\sin\theta\sin\omega t \qquad (2.28)$$

$$x = A\cos(\omega t - \theta) \qquad (2.29)$$

$$A = \sqrt{x_0^2 + (\frac{\dot{x}_0}{\omega})^2} \qquad (2.30)$$

$$tan\theta = \frac{\dot{x}_0}{\omega x_0} \qquad (2.31)$$

Single Degree of Freedom Systems

2.2.2 DAMPED FREE VIBRATIONS

Shown in Fig. 2.3 below is a damped one degree of freedom system with damping constant c (damped SDF) (Church, A. H., 1957).

$$F(t) = 0 \tag{2.32}$$

Equation of motion (Fig. 2.4)

$$m\ddot{x} + c\dot{x} + kx = 0 \tag{2.33}$$

$$x(0) = x_0 \tag{2.34}$$

$$\dot{x}(0) = \dot{x}_0 \tag{2.35}$$

Again here general solution is of the form Ae^{st}, because considering either $A\sin\omega t$ or $B\cos\omega t$ or even $A\sin\omega t + B\cos\omega t$ as earlier done for undamped SDF cannot be a general solution because of the middle term $c\dot{x}$, due to damping. It will not enable us to cancel out the time dependent terms $A\sin\omega t$ or $B\cos\omega t$

$$\therefore x = Ae^{st} \tag{2.36}$$

$$\dot{x} = Ase^{st} \tag{2.37}$$

$$\ddot{x} = As^2 e^{st} \tag{2.38}$$

Substituting the above in the equation of motion (2.33) we have

$$Ams^2 e^{st} + Ase^{st} + Ake^{st} = 0 \tag{2.39}$$

$$ms^2 + cs + k = 0 \tag{2.40}$$

$$s_{1,2} = -\frac{c}{2m} \pm \frac{1}{2}\sqrt{\frac{c^2}{m^2} - 4\omega^2} \tag{2.41}$$

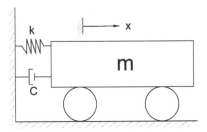

FIGURE 2.3: Damped SDF system

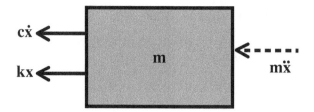

FIGURE 2.4: Free body diagram

Here $\frac{k}{m} = \omega^2$

$$s_1 = -\frac{c}{2m} - \frac{1}{2}\sqrt{\frac{c^2}{m^2} - 4\omega^2} \tag{2.42}$$

$$s_2 = -\frac{c}{2m} + \frac{1}{2}\sqrt{\frac{c^2}{m^2} - 4\omega^2} \tag{2.43}$$

There are only 3 cases possible:
Case(1): Critically damped ($c = c_{cr}$)

$$\frac{c^2}{m^2} = 4\omega^2 \tag{2.44}$$

$c = 2m\omega$ then the terms inside the square root cancel each other.

$$s_1 = s_2 = -\frac{c}{2m} \quad \ldots\ldots\ldots Repeated\ roots \tag{2.45}$$

The motion is not oscillatory. It comes to a halt after initial excitation. Such a system is called a "critically damped" one. Denoting c as c_{cr}

$$\frac{c_{cr}^2}{m^2} = 4\omega^2 \tag{2.46}$$

$$c_{cr} = 2m\omega \tag{2.47}$$

$$= 2m\sqrt{\frac{k}{m}} \tag{2.48}$$

$$= 2\sqrt{km} \tag{2.49}$$

$$x_1(t) = A_1 e^{-t(c_{cr}/2m)}, for\ s_1 = -\frac{c_{cr}}{2m} \tag{2.50}$$

Single Degree of Freedom Systems

$$x_2(t) = A_2 e^{-t(c_{cr}/2m)}, \text{ for } s_2 = -\frac{c_{cr}}{2m} \qquad (2.51)$$

As it is a case of repeated roots,

$$x(t) = (A_1 + A_2 t) e^{-t(c_{cr}/2m)} \qquad (2.52)$$

Case(2): Over damped ($c > c_{cr}$)

$c > 2m\omega$ or $c > c_{cr}$; it is over damped.

The roots $s_{1,2}$ are real. Here also, like the critically damped case, the system comes to a halt after some initial excitation. In both the cases, the system does not vibrate. However, surprisingly a critically damped system comes to a halt faster than an over damped system. Ref. Fig 2.5 for the three initial conditions.

Case(3): Under damped system ($c < c_{cr}$).

$$\frac{c^2}{m^2} - 4\omega^2 < 0 \qquad (2.53)$$

$$s_1 \neq s_2 \qquad (2.54)$$

s_1 and s_2 are complex conjugates, because the quantity under the square root is negative.

$$s_{1,2} = -\eta\omega \pm \sqrt{\eta^2\omega^2 - \omega^2} \qquad (2.55)$$

$$= -\eta\omega \pm i\omega\sqrt{1-\eta^2} \qquad (2.56)$$

$$= -\eta\omega \pm i\omega_d \qquad (2.57)$$

Where, $\omega_d = \omega\sqrt{1-\eta^2}$, damped natural frequency.
The general solution is,

$$x(t) = Ae^{s_1 t} + Be^{s_2 t} \qquad (2.58)$$

$$= Ae^{(-\omega\eta + i\omega_d)t} + Be^{(-\omega\eta - i\omega_d)t} \qquad (2.59)$$

$$= e^{-\omega\eta t}[Ae^{i\omega_d t} + Be^{-i\omega_d t}] \qquad (2.60)$$

$$= e^{-\omega\eta t}[A(\cos\omega_d t + i\sin\omega_d t) + B(\cos\omega_d t - i\sin\omega_d t)] \qquad (2.61)$$

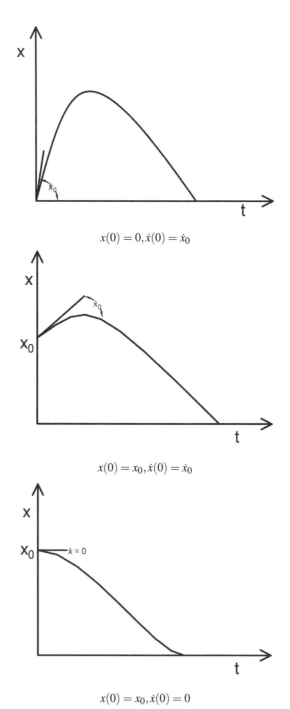

FIGURE 2.5: x vs. t for different initial conditions of $c = c_{cr}$ and $c > c_{cr}$

Single Degree of Freedom Systems

$$= e^{-\omega \eta t}[(A+B)\cos\omega_d t + i(A-B)\sin\omega_d t] \tag{2.62}$$

$$= e^{-\omega \eta t}[A_1\cos\omega_d t + A_2\sin\omega_d t] \tag{2.63}$$

$$= e^{-(\frac{c}{2m})t}[A_1\cos\omega_d t + A_2\sin\omega_d t] \tag{2.64}$$

ω_d is also,

$$\omega_d = \sqrt{\frac{k}{m} - (\frac{c}{2m})^2} \tag{2.65}$$

or

$$\omega_d = \omega\sqrt{1-(\eta)^2}, \quad \eta = \frac{c}{c_{cr}} \tag{2.66}$$

$$\omega = \sqrt{\frac{k}{m}} \tag{2.67}$$

If the initial conditions at t = 0 are known, $x(0) = x_0, \dot{x}(0) = \dot{x}_0$

$$x(t) = e^{-\eta \omega t}[x_0\cos\omega_d t + \frac{\dot{x}_0 + x_0\eta\omega}{\omega_d}\sin\omega_d t] \tag{2.68}$$

$$x(t) = A_3 e^{-\eta \omega t}\cos(\omega_d t - \beta) \tag{2.69}$$

$$A_3 = \sqrt{x_0^2 + (\frac{\dot{x}_0 + x_0\eta\omega}{\omega_d})^2} \tag{2.70}$$

$$\tan\beta = \frac{\dot{x}_0 + x_0\eta\omega}{\omega_d x_0} \tag{2.71}$$

Also, we note that

$$\omega \approx \omega_d \tag{2.72}$$

Because even for $\eta = 10\%$ or 0.1

$$\omega_d = \omega\sqrt{1-(0.1)^2}$$

$$= \omega\sqrt{1-0.01}$$

$$= \omega\sqrt{0.99} = 0.994\omega$$

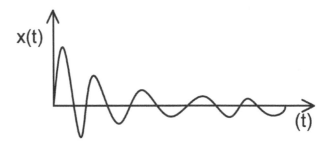

FIGURE 2.6: Plot of x(t) vs. t

2.2.3 LOGARITHMIC DECREMENT

Here we discuss about the logarithmic decrement in the case of viscously damped systems and here, it is a simple single degree of freedom system (Thomson, W. T., 1972) (Ref. Figs. 2.6 and 2.7). It may be appropriate to mention that there are different types of damping systems, i.e., viscous, frictional or coulomb, aerodynamic and so on. In all our structural modelling, we adopt viscous damping, as it is easy to model. Viscous damping, as the name implies, is damping one experiences when a plunger is thrust into a viscous medium. It is observed that the force of resistance offered is proportional to the velocity of the plunger entering the viscous fluid. The constant of proportionality is the damping constant c. The damping force, $F_e = c\dot{x}$ where $\dot{x}$ is the velocity of the plunger. It is known that the vibrations of a viscously damped system gradually decay exponentially. The exponent is a linear function of the damping factor η. If x_1 and x_2 are two amplitudes one cycle apart at times t_1 and t_2, respectively, the ratio of the amplitudes is

$$\frac{x_1}{x_2} = \frac{\bar{x}_1 e^{-\eta \omega t_1} \cos(\omega_d t_1 - \Phi)}{\bar{x}_2 e^{-\eta \omega t_2} \cos(\omega_d t_2 - \Phi)}$$

$t_2 = t_1 + T$, where T is the period $\frac{2\pi}{\omega_d}$ is the period of damped oscillation.

$$\eta = \frac{c}{c_c}$$

Also,

$$\cos(\omega_d t_2 - \Phi) = \cos[(\omega_d t_1 - \Phi) + \omega_d T])$$
$$= \cos[(\omega_d t_1 - \Phi) + 2\pi)]$$
$$= \cos[(\omega_d t_1 - \Phi)]$$

$$\frac{x_1}{x_2} = \frac{e^{-\eta \omega t_1}}{e^{-\eta \omega (t_1 + T)}} = e^{\eta \omega T}$$

$$\eta \omega t = \ln \frac{x_1}{x_2} = \frac{2\pi \eta}{\sqrt{1 - \eta^2}}$$

$\ln \frac{x_1}{x_2}$, denoted by δ is called the "logarithmic decrement."

Single Degree of Freedom Systems

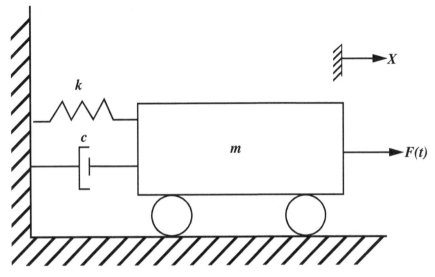

FIGURE 2.7: Forced vibration SDF system

2.3 FORCED VIBRATION OF A DAMPED SINGLE DEGREE OF FREEDOM

Let F(t) be a harmonic force generally defined as $F(t) = \overline{F}e^{i\Omega t}$ where Ω = Forcing frequency (Craig, Roy R., 1981). Let us remember

$$e^{i\Omega t} = cos\Omega t + isin\Omega t \tag{2.73}$$

$$m\ddot{x}(t) + c\dot{x}(t) + kx(t) = \overline{F}e^{i\Omega t} \tag{2.74}$$

The general solution of the above differential equation can be always written as complimentary function + particular integral.
The particular solution lies in the form of $Ae^{i\Omega t}$ where A is any constant.
The complimentary function can be of the form of $A_1 sin\omega t + B_1 cos\omega t$ where ω is the natural frequency and A_1, B_1 are constants to be found from initial conditions.

Substituting the particular integral, $x_{PI} = Ae^{i\Omega t}$ in Equation (2.76)

$$-m\Omega^2 A + ic A\Omega + kA = \overline{F} \tag{2.75}$$

From the above Equation (2.75)

$$A = \frac{\overline{F}}{k - m\Omega^2 + ic\Omega} \tag{2.76}$$

$$\therefore x_{PI} = \frac{\overline{F}e^{i\Omega t}}{k - m\Omega^2 + ic\Omega} \tag{2.77}$$

It is better to eliminate the imaginary number i in the denominator. Therefore, we make use of the well-known identity.

$$e^{i\alpha} = cos\alpha + isin\alpha \quad (2.78)$$

write,

$$x_{PI} = \frac{\overline{F}e^{i\Omega t}}{(\sqrt{(k-m\Omega^2)^2 + (c\Omega)^2})e^{i\alpha}} \quad (2.79)$$

$$x_{PI} = \frac{\overline{F}e^{i(\Omega t - \alpha)}}{\sqrt{(k-m\Omega^2)^2 + (c\Omega)^2}} \quad (2.80)$$

and $tan\alpha = \frac{c\Omega}{k-m\Omega^2}$

If the force component is $sin\Omega t$, then

$$x_{PI} = \frac{\overline{F}sin(\Omega t - \alpha)}{\sqrt{(k-m\Omega^2)^2 + (c\Omega)^2}} \quad (2.81)$$

Or

$$x_{PI} = A_3 sin(\Omega t - \alpha) \quad (2.82)$$

Where the amplitude A_3 is

$$A_3 = \frac{\overline{F}}{\sqrt{(k-m\Omega^2)^2 + (c\Omega)^2}} \quad (2.83)$$

also

$$x_{PI} = \frac{x_{st} sin(\Omega t - \alpha)}{\sqrt{(1-r^2)^2 + (2\eta r)^2}} \quad (2.84)$$

$$tan\alpha = \frac{2\eta r}{1 - r^2} \quad (2.85)$$

where $x_{st} = \frac{\overline{F}}{k}$ is the static deflection $\eta = \frac{c}{c_c}$

$r = \frac{\Omega}{\omega}$, r = Ratio of frequencies $= \frac{forcing\ frequency}{natural\ frequency}$

The general solution now is = Complimentary part + Particular integral

$$x(t) = e^{-\eta \omega t}[(A_1 sin(\omega_d t) + B_1 cos(\omega_d t) + \frac{x_{st} sin(\Omega t - \alpha)}{\sqrt{(1-r^2)^2 + (2r\eta)^2}}] \quad (2.86)$$

A_1 and B_1, are found from the initial conditions.

Single Degree of Freedom Systems

The first two terms are due to initial conditions and therefore contain the natural frequency, while the 3rd term is due to the force, and therefore contains the forcing frequency Ω.

Although Equation (2.87) for x(t) is a complete solution, eventually with time, the vibration due to initial conditions die out due to damping or more precisely due to the effect of multiplication factor $e^{-\eta \omega t}$, where the negative exponent is the cause for exponential decay. Therefore, what remains, as steady state under the influence of the force is the 3rd term.

It is,

$$\text{Steady state } X(t) = \frac{X_{st} \sin(\Omega t - \alpha)}{\sqrt{(1-r^2)^2 + (2r\eta)^2}} \quad (2.87)$$

The above Equation (2.87) for X(t) is a time dependent function, the amplitude of which is

Amplitude $= \frac{X_{st}}{\sqrt{(1-r^2)^2 + (2\eta r)^2}}$.

On dividing the amplitude by X_{st} to obtain a non-dimensional ratio, we get what is termed as magnification factor (MF) also called dynamic magnification factor or dynamic load factor (DLF).

MF or DLF $= \frac{1}{\sqrt{(1-r^2)^2 + (2\eta r)^2}}$. It is plotted in Fig. 2.8.

The plot of MF or DLF versus $r = \frac{\Omega}{\omega}$, Fig. 2.8 shows the plot. The various plots are for various values of η. As η increases, the peaks reduce and also gradually shift to the left towards the origin. They never occur at r = 1, but towards slightly left of r =1. Only when $\eta = 0$, the imaginary peak, the point of resonance occurs at r = 1, where magnification factor or DLF is infinity. The peak occurs at r = $\sqrt{(1-2\eta^2)}$.

The DLF or MF is also the magnitude of the frequency response where frequency response or complex frequency response is denoted as

$$G(i\Omega) = \frac{1}{1 - r^2 + i2\eta r} \quad (2.88)$$

There is also what is called quality factor Q (term borrowed from electrical engineering), which is the maximum value of DLF or MF, in the region around r = 1 (the so-called point of resonance), for small values of η, $Q \approx \frac{1}{2\eta}$.
At two points over which the difference of frequencies $\Omega_1 - \Omega_2$, which is also called the the (half-power) bandwidth, the value of DLF forcing or MF will be $\frac{Q}{\sqrt{2}}$. The term is so called because the power consumed by an electric circuit at those points or by a damper in a vibrating system will be proportional to the square of the amplitude, of course when the system is harmonically vibrating. The difference in frequency ratios $r_2 - r_1 = \frac{\Omega_2 - \Omega_1}{\omega} \approx 2\eta$ and therefore, $\Omega_2 - \Omega_1 = 2\eta \omega$ or (half-power) bandwidth = $2\eta \omega$

$$\therefore Q = \frac{1}{2\eta} \approx \frac{\omega}{\Omega_2 - \Omega_1} \quad (2.89)$$

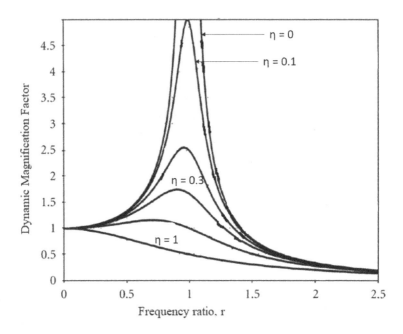

FIGURE 2.8: Dynamic magnification factor as a function of frequency ratio for various values of damping

Explaining the frequency response, the fact is when the natural frequency is large compared to the forcing frequency or when the system is highly rigid, the system does not respond dynamically. Its response will be static (extreme left of Fig. 2.8).

On the other hand, when the system is very flexible its natural frequency is very low, and relatively the forcing frequency is very high, the system does not respond at all. Its response will be almost zero as you observe towards extreme right of the Fig.2.8. Only when the system's natural frequency is close to the forcing frequency the response will be large and resonance occurs. Without damping, the response will be infinitely large, which does not happen in practice, because of the presence of damping.

The author always chose to give examples of everyday life comprising a similar phenomenon in vibration problems to drive the fact much better and deeper.

In our daily life, let us take the example of a boss and an employee. If the employee is fast in his reaction (high frequency) and relatively the boss is quite slow (low frequency), there will not be much output. The output of the employee will be simply static on the other hand, if the employee is too slow (very flexible) compared to the boss who is very fast, then the employee will not be able to cope and will not be able to work at all. Only, when the boss and employee are close to each other in their efficiency, then only the output will be optimum.

2.4 A SINGLE DEGREE OF FREEDOM SYSTEM SUBJECTED TO SUPPORT MOTION

We know that in rigid systems (when the natural frequency is very high), the acceleration of the mass is the same as that of the support or base (acceleration sensitive), while in very flexible systems (where the natural frequency is very low) the mass hardly moves (Mario Paz, 2004). Let us take the example of a short rigid bar fixed vertically on a table. If the table moves, the bar being very rigid moves along with the table. Instead, if it is a very thin flexible stick, the table moves, but the flexible stick bends with its top just stationary.

Now let us see how a one degree of freedom system behaves when subjected to support motion. The support motion can be either time dependent displacement or acceleration. Shown in Fig. 2.9 is a typical one-storey building frame to be designed for earthquakes. Before the design, obviously, we need to analyze the frame for the design earthquake. As we know, an earthquake is a ground motion. The ground motion can be expressed in terms of time dependent displacement, velocity or acceleration. In any recorded earthquake ground motion, as already mentioned, there will be time history of ground displacement, velocity and acceleration in three directions: two horizontal and one vertical. However, it is very convenient to analyze and design when earthquake ground accelerations are known. Therefore, we shall discuss the response of a one degree of freedom system subjected to support acceleration.

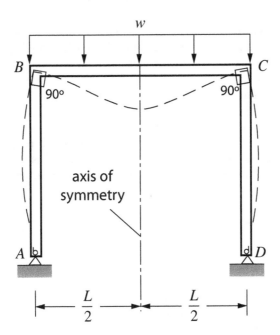

FIGURE 2.9: Typical one-storey building

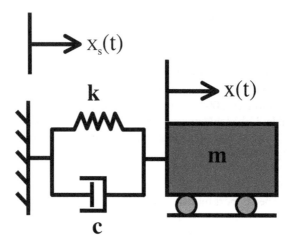

FIGURE 2.10: SDF system

Generally shear building approximation is adopted for simplicity. With the advent of modern softwares, like ETABS, STAAD PRO and ANSYS, such a simplification is not necessary. Finite element idealization is adopted.

In the shear building approximation, the entire mass is lumped at the floor/roof level of the frame. The columns are considered to resist the lateral forces due to earthquakes. The total mass to be lumped consists of dead and live loads. The self-weight of the columns and infills are also lumped at the floor level shared by top and bottom floors. The equivalent model which represents the frame is a simple mass representing the mass of the frame and spring representing the columns. The lateral stiffness of the columns only constitutes the spring stiffness k.

The equivalent spring mass system is shown in Fig. 2.10.

It is a one degree of freedom system with x as the degree of freedom, g = acceleration due to gravity, k = spring stiffness = $2 \times \frac{12EI}{h^3}$ where $\frac{12EI}{h^3}$ is the horizontal shear force offered by each of the columns when its end is displaced, similar to that of the beam in Fig. 2.11.

Now let us write the equation of motion,

$$m\ddot{x}_r + c\dot{x}_r + kx_r = 0$$

The spring force depends on the relative displacement $x_r = x - x_s$. The right-hand side is zero because there is no external force on the mass. Because the spring force and the damping force contain x_r, the first term, namely inertia force, also should contain x_r. Expressing x in terms of x_r,

Single Degree of Freedom Systems

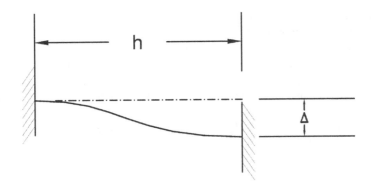

FIGURE 2.11: A member of span h is subjected to support displacement Δ

$$x = x_r + x_s$$
$$m\ddot{x}_r + c\dot{x}_r + kx_r = -m\ddot{x}_s$$

$\ddot{x}_s$ is the support acceleration. It is the earthquake ground acceleration. Let us simply assume that $x_s = \bar{x}_s \sin\Omega t$. In fact, earthquake ground acceleration is a kind of random sinusoidal motion.

Then,

$$m\ddot{x}_r + c\dot{x}_r + kx_r = -m\bar{x}_s\Omega^2 \sin\Omega t$$

$$= F\sin\Omega t$$

$$Where, F = -m\bar{x}_s\Omega^2$$

The support motion is in fact equivalent to a force applied on the mass of the form where F is the amplitude of the force. The equation is similar to the equation of motion in forced vibration which we have already seen. It can be easily solved either by classical methods or numerically, if the force is random as in earthquake ground acceleration.

2.5 RAYLEIGH'S METHOD TO OBTAIN NATURAL FREQUENCY

Rayleigh's method has an important place in the theory of vibration as it demonstrates the conservation of energy. Of course, as the name implies, the energy conservation can take place only when there is neither loss nor dissipation of energy. Even addition of energy to the system should be absent. Therefore, it cannot be valid for a system with damping because the damping dissipates energy, and also it cannot be valid for forced vibration problems because there is addition of energy. It can be valid only for an undamped freely vibrating system which means the usefulness is only in obtaining natural frequency. The principle of conservation of energy is the total energy of any point of time of a freely vibrating undamped system is constant.

Total energy (E) is the sum of kinetic energy (KE) and potential energy (PE). The potential energy (PE) = (1/2) k x^2. The kinetic energy (KE) = (1/2) m $\dot{x}^2$. The sum is

$$E = (\frac{1}{2})kx^2 + (\frac{1}{2})m\dot{x}^2 = constant \qquad (2.90)$$

$$\frac{dE}{dt} = \frac{1}{2}(2m\dot{x}\frac{d\dot{x}}{dt}) + \frac{1}{2}(2kx\frac{dx}{dt}) = 0 \qquad (2.91)$$

$$m\ddot{x}\dot{x} + kx\dot{x} = 0 \qquad (2.92)$$

or

$$m\ddot{x} + kx = 0 \qquad (2.93)$$

The above equation is a well-known equation of motion. Let us see how we obtain the natural frequency. As the total energy E is always constant, it should be true at the maximum value also. But the advantage we have in considering E_{max} as constant is that we can find either kinetic energy zero when the potential energy is maximum or potential energy zero when the kinetic energy is maximum. Because it is natural that when the displacement x is at its maximum, it will be about to reverse and come backwards, and therefore the velocity $\dot{x} = 0$, as it has to change its sign or the direction at that instant.

Similarly, when the mass has returned to its equilibrium position from the maximum displacement x_m, its velocity $\dot{x}$ should be maximum going in the opposite direction. See Fig. 2.12 below.

Therefore, Max. KE + Max. PE = constant.
∴ 0 + Max. PE = constant.
or Max. KE + 0 = constant.
Max. KE is (1/2) m $\dot{x}_m^2$.
Max. PE is = (1/2) k x_m^2.

Consider a simple harmonic oscillation,

$$x = \bar{x}\sin\omega t \qquad (2.94)$$

Single Degree of Freedom Systems

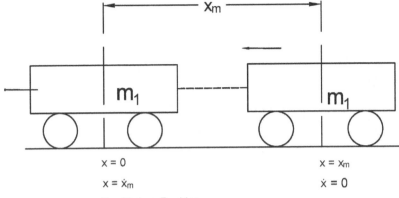

FIGURE 2.12: SDF system at extreme positions

$$x_m = \bar{x}, \dot{x}_m = \bar{x}\omega \tag{2.95}$$

$$Max.P.E = \frac{1}{2}k\bar{x}^2 \tag{2.96}$$

$$Max.K.E = \frac{1}{2}m(\bar{x}\omega)^2 \tag{2.97}$$

$$\frac{1}{2}k\bar{x}^2 = \frac{1}{2}m(\bar{x}\omega)^2 \tag{2.98}$$

$$\omega^2 = \frac{k}{m} \tag{2.99}$$

$$\omega = \sqrt{\frac{k}{m}} \tag{2.100}$$

2.6 RESPONSE IN FREQUENCY DOMAIN AND LAPLACE TRANSFORMATION

We have understood the time domain response, which means we write both the equations of motion and the resulting response to a time varying input in terms of the real time only (Leonard Meirovitch, 1986). Very often it will be convenient and elegant to write both the system equations and the input in terms of frequencies (both the

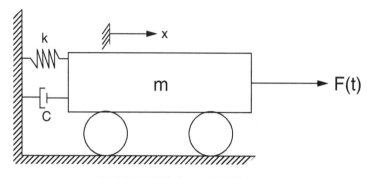

FIGURE 2.13: Damped SDF system

natural frequency and the forcing or input frequency Ω) and obtain the response also obviously in terms of the frequencies. For the above, we need to write the equations of motion in time domain and then write them in terms of Laplace variable **s**, and perform a simple transformation called "Laplace transformation." Before we proceed with the Laplace transformation, let us see how we get the complex response in time domain. We illustrate the method with respect to a SDF for simplicity (Fig. 1.2).

The equation of motion for SDF in time domain is,

$$m\ddot{x}(t) + c\dot{x}(t) + kx(t) = F(t) \tag{2.101}$$

Dividing by 'm'

$$\ddot{x}(t) + \frac{c}{m}\dot{x}(t) + \frac{k}{m}x(t) = \frac{F}{m}(t) \tag{2.102}$$

For convenience of writing, we can drop 't' within the brackets, although it will mean that both 'x' and 'F' are functions of 't'.

$$\frac{c}{m} = \frac{c}{c_{cr}} \cdot \frac{c_{cr}}{m} = \eta \frac{2\sqrt{km}}{m} \tag{2.103}$$

Where, $\eta = \frac{c}{c_{cr}}$ is a non-dimensional factor called "damping factor."

$$= 2\eta \sqrt{\frac{k}{m}} \tag{2.104}$$

$$= 2\eta \omega \tag{2.105}$$

Let $F(t) = \overline{F} f(t)$
where, $\overline{F}$ is the amplitude of force and $f(t)$ is the time function.

$$\ddot{x} + 2\eta \omega \dot{x} + \omega^2 x = \frac{\overline{F}}{m} f(t) \tag{2.106}$$

Single Degree of Freedom Systems

$$= \frac{\omega^2 \overline{F}}{k} f(t) \qquad (2.107)$$

Let x in steady state be assumed in the general complex form as $x = \overline{x}e^{i\Omega t}$. Where, $f(t) = e^{i\Omega t}$,
$\overline{x}$ is amplitude of displacement.

Substituting x in (2.108), as

$$\text{Where,} \quad \dot{x} = (i\Omega)\overline{x}e^{i\Omega t} \qquad (2.108)$$

$$\ddot{x} = (i\Omega)^2 \overline{x}e^{i\Omega t} = -\Omega^2 \overline{x}e^{i\Omega t} \qquad (2.109)$$

$$(-\Omega^2 \overline{x} + 2\eta \omega i \Omega \overline{x} + \omega^2 \overline{x})e^{i\Omega t} = (\omega^2 x_{st})e^{i\Omega t} \qquad (2.110)$$

$\because \overline{F} = kx_{st}$
and $\frac{k}{m} = \omega^2$
x_{st} is a constant in displacement units and also a function of $i\Omega$ and can be written as $\overline{x}(i\Omega)$

$$\overline{x}[\omega^2 - \Omega^2 + 2\eta i \Omega \omega]e^{i\Omega t} = x_{st}(\omega^2 e^{i\Omega t}) \qquad (2.111)$$

We can observe that, both the natural frequency ω and input frequency Ω are involved along with 't' of course, which is the independent variable.

If $Z(i\Omega)$ is chosen to represent $[\omega^2 - \Omega^2 + 2\eta i \Omega \omega]$, called the impedance function

$$\overline{x}Z(i\Omega)e^{i\Omega t} = x_{st}\omega^2 e^{i\Omega t} \qquad (2.112)$$

Then,

$$\overline{x}Z(i\Omega) = x_{st}\omega^2 \qquad (2.113)$$

$$\overline{x} = \frac{x_{st}}{Z(i\Omega)}\omega^2 \qquad (2.114)$$

Taking ω^2 to the denominator

$$\overline{x} = \frac{x_{st}}{[1 - \frac{\Omega^2}{\omega^2} + 2\eta \frac{i\Omega}{\omega}]}$$

$G(i\Omega)$ denotes what is defined as frequency response in the ratio of $\frac{\overline{x}}{x_{st}} = \frac{1}{(1-(\frac{\Omega}{\omega})^2) + 2\eta \frac{i\Omega}{\omega}}$

$G(i\Omega) = \frac{\overline{x}}{x_{st}}$

The magnitude $|G(i\Omega)| = \dfrac{1}{([(1-(\frac{\Omega}{\omega})^2)]+[2\eta\frac{\Omega}{\omega}]^2)^{\frac{1}{2}}}$ is also known as the magnification factor.

Coming back to time domain,

$$x(t) = x_{st} G(i\Omega) e^{i\Omega t}$$

$$= x_{st} |G(i\Omega)| e^{i(\Omega t - \theta)}$$

where θ is the phase angle.

Now let us study the Laplace transformation method for SDF.

LAPLACE TRANSFORMATION METHOD FOR SDF SYSTEMS

Laplace transformation is an elegant method by which we transform or convert a differential equation into an algebraic equation. We are all familiar with the solution of an algebraic equation, like for example a quadratic equation. It is simplest to obtain a solution, as we have all learnt it from the high school level. After we obtain the solution for the algebraic equation, we get back to its original solution of the differential equation which was really what was required by an inverse transformation. Laplace transform and its inverse transformation are readily available in an equally simple way. We may note that while the variable real time t appears in the differential equation after transforming it to an algebraic equation by applying Laplace transformation a new variable 's' appears in place of 't'. Laplace transformation of a function $x(t)$ is denoted by $Lx(t) = x(s)$. After transformation $x(t)$ becomes $x(s)$. To obtain $x(t)$ from $x(s)$ we perform what is called "inverse transformation." $x(t) = L^{-1} x(s)$.

Laplace transformation is denoted by Lx(t)

$$L(x(t)) = \int_0^\infty e^{-st} x(t) dt$$

is defined as Laplace transform of the function x(t). Because the variable of integration is 't', after applying the limits the expression contains the variables 's', which is in its most general form a complex variable. By applying the operator Lx(t) on any differential equation, we convert or transform a differential equation to algebraic equation in 's' only. While the differential equation is in the time domain, the transformed algebraic equation will be in 's' domain or in other words a complex plane. Because Lx(t) will be function of 's' as we consider it as x(s).

After defining Lx(t), let us define $L\dfrac{dx(t)}{dt}$

$$L\dfrac{dx(t)}{dt} = \int_0^\infty e^{-st} \left(\dfrac{dx(t)}{dt}\right) dt$$

After integration by parts we get $e^{-st} x |_0^\infty + s \int_0^\infty e^{-st} x \, dt$

Single Degree of Freedom Systems

The initial displacement and velocity as t=0 can be denoted by $x_0, \dot{x}_0$. The above equation now can be written as

$$= 0 - x_0 + sLx(t)$$

$$= 0 - x_0 + sx(s)$$

$$= sx(s) - x_0$$

Similarly $L\frac{d^2x}{dt^2} = \int_0^\infty e^{-st} \frac{d^2x}{dt^2} dt$

$$= s^2 x(s) - sx_0 - \dot{x}_0$$

Further we also need L of the excitation force because we will be applying the operator L throughout the equation of motion.

$LF(t) = \int_0^\infty e^{-st} F(t) dt$ can be denoted as $F(s)$. Now applying the operator L on the equation of motion of a SDF, which of course will be in the time domain, or in other words, we will be taking the Laplace transform of the equation of motion

$$L[m\frac{d^2x}{dt^2} + c\frac{dx}{dr} + kx] = LF(t)$$

Transforming term by term

$$mL\frac{d^2x}{dt^2} + cL\frac{dx}{dr} + kLx = LF(t)$$

$$(ms^2 + cs + k)x(s) = F(s) + m\dot{x}_0 + (ms + c)x_0$$

It is convenient to assume that both $\dot{x}_0 = x_0 = 0$ so that the understanding is easier

$$\therefore (ms^2 + cs + k)x(s) = F(s)$$

Now defining Z(s) as the generalized impedance which is also the transformed impedance,

$$Z(s) = \frac{F(s)}{x(s)} = ms^2 + cs + k$$

Z(s) is obviously an algebraic expression in s.

Now another function called "transfer function" is very commonly employed in dynamics. It is defined as the ratio of transformed output to the transformed input. It is generally denoted as G(s)

and $G(s) = \frac{x(s)}{F(s)}$

All transformed quantities will be a function of 's', the Laplace variable, while all quantities before transformation will be a function of time. We can also say that a differential equation, a function of time, will become an algebraic equation as a function of 's', the Laplace variable.

To get back the previous function x(t), we have to perform what is called inverse transformation.

$$x(t) = L^{-1}x(s)$$

We shall not go very deep into this subject, as they have been dealt with in great detail in other books. However, certain important conclusions like the following will be useful.

If $\delta(t)$ represents unit impulse, then Laplace transform L of $\delta(t)$ is
$L\delta(t) = \delta(s) = \int_0^\infty e^{-st}\delta(t)dt = 1$

Similarly, Laplace transform of a unit step function is,

$$f(s) = \int_0^\infty e^{-st} f(t)dt$$
$$= \int_0^\infty e^{-st} dt$$
$$= \frac{1}{s}$$

Solution of an algebraic equation is easier than that of differential equation and therefore transforming the differential equation to an algebraic equation is better as solution is simpler.

2.7 SOLVED PROBLEMS

(1) A harmonic motion $x = \bar{x}\sin\omega t$ has a time period of 0.2 sec and an amplitude of 0.8 cm. Find maximum velocity and acceleration.

Given: T = 0.2, $\bar{x}$ = 0.8 cm

Frequency = $\omega_n = \frac{2\pi}{T} = \frac{2\pi}{0.2} = 31.42$ rad/sec

Maximum velocity = $\bar{x}\omega = 0.8 \times 31.42 = 25.13$ cm/sec

Maximum acceleration = $\bar{x}\omega^2 = 0.8 \times 31.42^2 = 789.77$ cm/sec^2

(2) A harmonic motion $x = \bar{x}\sin\omega t$ has maximum velocity of 4 m/s and has frequency of 12 cycles per sec. Determine its amplitude, period and maximum acceleration.

Single Degree of Freedom Systems

Given:

Frequency = 12 cps = $\frac{1}{T} = \frac{\omega}{2\pi}$

$\omega = 2\pi \times 12 = 24\pi = 75.398$ rad/sec

$T = \frac{2\pi}{\omega} = \frac{2\pi}{75.398} = 0.0833$ sec

Maximum velocity = $\bar{x}\omega$

$4 = \bar{x}(75.398)$

$\bar{x} = 0.053$ m

$\bar{x} = 53$ mm

Maximum acceleration = $\bar{x}\omega^2$

$= 53 \,(75.398)^2 = 301297.5$ mm/sec

(3) A one kg mass is suspended by a spring having a stiffness of 5 N/mm. Determine the natural frequency and static deflection of the spring.

Given: k = 5 N/mm = 5000 N/m

m = 1 kg

$\omega = \sqrt{\frac{k}{m}} = \sqrt{\frac{5000}{1}} = 70.71$ rad/sec

$f = \frac{\omega}{2\pi} = \frac{70.71}{2\pi} = 11.25\ Hz$

$\omega^2 = \frac{9.81}{\delta}$

$\delta = \frac{9.81}{70.71^2} = 0.00196\ m$

$= 1.96$ mm

(4) A vertical cantilever beam of 3 m long supports a mass of 300 kg at the top. Find natural period and natural frequency. Take $E = 2.1 \times 10^5$ N/mm^2 and I = 1200 cm^2.

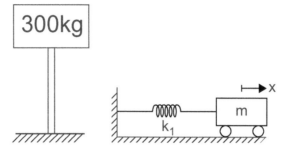

FIGURE 2.14: Vertical cantilever beam with the mass and the corresponding spring mass system

Given:

Cantilever beam span = 3 m

Assuming the rotation of column at the mass is allowed.

$k = \frac{3EI}{L^3}$

$= \frac{3 \times 2.1 \times 10^5 \times 1200 \times 10^4}{3000^3}$

$= 280 \times 1000$ N/m

$\omega = \sqrt{\frac{280 \times 1000}{300}} = 30.55$ rad/sec

$f = \frac{\omega}{2\pi} = \frac{30.55}{2\pi} = 4.86$ Hz

$T = \frac{1}{f} = 0.21$ sec

Sometimes the mass could be so large and rigid, in which case it prevents the column rotations

$k = \frac{12EI}{L^3} = 1120 \times 10^3$ N/m.

$\omega = \sqrt{\frac{k}{m}} = \sqrt{\frac{1120 \times 10^3}{300}} = 61.1$ rad/sec

$f = \frac{\omega}{2\pi} = 9.72$ Hz

$T = \frac{1}{f} = 0.102$ sec

Single Degree of Freedom Systems

(5) Find the natural frequency of the system as shown in Fig. 2.15 below.

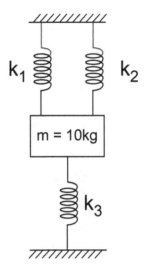

FIGURE 2.15: Spring-Mass system

Given:

$k_1 = k_2 = 3000$ N/m

$k_3 = 1500$ N/m

m = 10 kg

As the springs are in parallel

$k_{e1} = k_1 + k_2$

$= 3000 + 3000 = 6000$ N/m

The equivalent spring is parallel to k_3

$k_e = k_{e1} + k_3$

$= 6000 + 1500 = 7500$ N/m

$\omega = \sqrt{\frac{k_e}{m}} = \sqrt{\frac{7500}{10}} = 27.386$ rad/sec

$f = \frac{\omega}{2\pi} = \frac{27.386}{2\pi} = 4.358$ Hz

(6) For the system shown in Fig. 2.16 below and the stiffness mentioned, find the mass of the spring if the system has a natural frequency of 20 Hz.

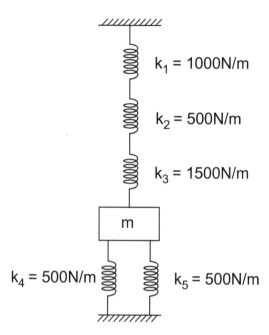

FIGURE 2.16: Spring-Mass system

k_1, k_2, k_3 are in series

$\frac{1}{k_{e1}} = \frac{1}{k_1} + \frac{1}{k_2} + \frac{1}{k_3} = \frac{1}{1000} + \frac{1}{500} + \frac{1}{1500}$

$k_{e1} = 272.72$ N/m

k_4 and k_5 are in parallel

$k_{e2} = 500 + 500 = 1000$ N/m

$k_e = k_{e1} + k_{e2} = 272.72 + 1000 = 1272.72$ N/m

$f = \frac{\omega}{2\pi} \Rightarrow \omega = 20 \times 2\pi = 125.66$ rad/sec

$\omega = \sqrt{\frac{k_e}{m}} \Rightarrow m = \omega^2 k_e = 125.66^2 \times 1272.72 = 20$ kN

(7) Obtain the differential equation of motion for the system shown in Figs. 2.17 and 2.18.

Single Degree of Freedom Systems 35

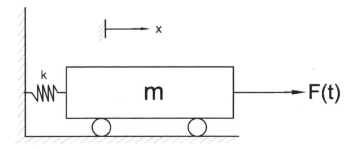

FIGURE 2.17: Single degree of freedom system

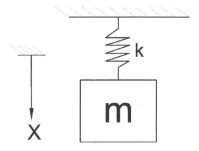

FIGURE 2.18: Single degree of freedom system

Solution:

Fig. 2.19 and Fig. 2.20 show the free body diagrams. (FBD) of Fig. 2.17 and Fig. 2.18, respectively.

In FBD 'm' represents mass of the oscillator displaced in y direction and acted upon spring force

$$F = k*x$$

The weight of the body, mg, and normal reaction, N, is also shown. Applying Newton's Law of Motion, thus,

$$m\ddot{x} + kx = 0$$

As shown in Fig. 2.20, when the spring is in the equilibrium position, the spring is stretched x_0 units and exerts force $kx_0 = W$ in the upward direction. W = weight of the body. When the body is displaced at a distance 'y' downward, the magnitude of the force in the spring is given as

$$F_s = k(x_0 + x)$$

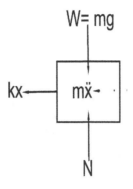

FIGURE 2.19: Free body forces

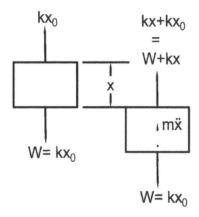

FIGURE 2.20: Free body forces

or

$$F_s = W + kx$$

Applying Newton's Law of Motion

$$-(W + kx) + W = m\ddot{x}$$

thus,

$$m\ddot{x} + kx = 0$$

(8) Determine the natural frequency of the system shown in Fig. 2.21 consisting of weight 250 N attached to a horizontal cantilever beam through the spring k_2.

Single Degree of Freedom Systems

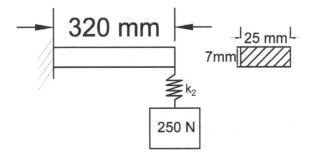

FIGURE 2.21: Take E = 2 × 10⁵ N/mm² and k_2 = 2 N/mm

Solution

The deflection at the free end of cantilever beam is given by

$$\Delta = \frac{PL^3}{3EI}$$

Corresponding spring constant

$$k_1 = \frac{P}{\Delta} = \frac{3EI}{L^3}$$

where, $I = \frac{bt^3}{12} = \frac{25 \times 7^3}{12} = 714.58$, for rectangular section. From the figure it is clear that both beam and spring are in series. The equivalent spring constant is

$$\frac{1}{k_e} = \frac{1}{k_1} + \frac{1}{k_2}$$

$$k_1 = \frac{3EI}{L^3} = \frac{3 \times 2 \times 10^5 \times 741.58}{320^3} = 13.57 \ N/mm$$

And k_2 = 2 N/mm, thus,

$$\frac{1}{k_e} = \frac{1}{13.57} + \frac{1}{2} = 0.57$$

and k_e = 1.74 N/mm =1.74 × 10³ N/m. The natural frequency is given by,

$$\omega = \sqrt{\frac{k_e}{m}}$$

$$\omega = \sqrt{\frac{1.74 \times 10^3 \times 9.81}{250}} = 8.26 \ rad/sec$$

Time period, $T = \frac{2\pi}{\omega} = \frac{2 \times 3.141}{8.26} = 0.76 \ sec$. Frequency, $f = T^{-1} = 1.31 \ Hz$

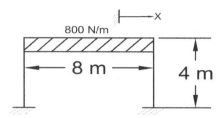

FIGURE 2.22: Rigid frame

(9) Consider a rigid steel frame as shown in Fig. 2.22 subjected to a dynamic force at upper level. Determine the natural frequency of the frame. Assume mass of columns is negligible, and the girder is rigid enough to prevent the rotation at the top of the columns. Take EI = 68.88 × 10^9 N-mm^2.

Solution

Here, W = 800 × 8 = 6400 N

$$k = \frac{12EI}{L^3} = \frac{12 \times 2 \times 68.66 \times 10^9}{4000^3} = 25.74 \, N/mm = 25.74 \times 10^3 \, N/m$$

The natural frequency is given by,

$$\omega = \sqrt{\frac{k_e}{m}}$$

$$\omega = \sqrt{\frac{25.74 \times 10^3 \times 9.81}{6400}} = 6.28 \, rad/sec$$

Time period, $T = \frac{2\pi}{\omega} = \frac{2 \times 3.141}{6.28} = 0.99 \, sec$

(10) For a system having a weight of 1 kN and stiffness k = 100 kN/m, amplitude of vibration decreases to 0.2 times the initial value, after 7 cycles, find damping coefficient c.

Solution:

Here $x_0 = 1$, $x_N = 0.2$, N = 7

$$\delta = \frac{1}{N}ln(\frac{x_0}{x_N}) = \frac{1}{7}ln(\frac{1}{0.2}) = 0.23$$

$$\eta = \frac{\delta}{\sqrt{\delta^2 + 4\pi^2}} = \frac{0.23}{\sqrt{0.23^2 + 4\pi^2}} = 0.0365$$

Single Degree of Freedom Systems

$$c = \eta c_{cr} = \eta \times 2 \times \sqrt{km} = 2 \times 0.0365 \times \sqrt{\frac{100 \times 10^3 \times 1000}{9.81}} = 233 \; N \; sec/m$$

(11) A simply supported beam carries a weight of 100 kN at center. The motor placed at the center of machine rotates at 350 rpm and unbalanced weight is 100 N with an eccentricity, e, of 0.25 m. If length of the beam is 4.75 m, modulus of elasticity, E, is 2×10^5 N/mm² and moment of inertia, I, is 50×10^6 mm⁴ and damping ratio, η, is 0.09. Find steady-state amplitude.

Solution:

Frequency of motor or forcing frequency,

$$\Omega = \frac{2\pi N}{60} = \frac{2 \times \pi \times 350}{60} = 36.64 \; rad/sec$$

Force transmitted by motor:

$$F = m_r e \omega^2 = \frac{100 \times 0.25 \times 36.64^2}{9.81} = 3421.22 \; N$$

Steady-state amplitude ($\overline{X}$) is given by

$$\overline{X} = \frac{F/k}{\sqrt{(1-r^2) + (2\eta r)^2}}$$

stiffness, k, for simply supported beam is given by

$$k = \frac{P}{\Delta} = \frac{48EI}{L^3} = \frac{48 \times 2 \times 10^5 \times 50 \times 10^6}{4750^3} = 4478.78 \; N/mm = 4.47 \times 10^6 \; N/m$$

Natural frequency of beam

$$\omega = \sqrt{\frac{k}{m}} = \sqrt{\frac{4.47 \times 10^6 \times 9.81}{100 \times 1000}} = 20.94 \; rad/sec$$

Frequency ratio r is given by

$$r = \frac{\Omega}{\omega} = \frac{36.64}{20.94} = 1.75$$

substituting all the values in equation, the steady-state amplitude is

$$\overline{X} = \frac{F/k}{\sqrt{(1-r^2)^2 + (2\eta r)^2}} = \frac{3421.22}{4.47 \times 10^6 \times \sqrt{(1-1.75^2)^2 + (2 \times 0.9 \times 1.75)^2}}$$

$$\overline{X} = 2.03 \times 10^{-4} \; m = 0.203 \; mm$$

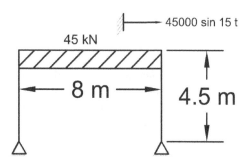

FIGURE 2.23: Steel frame

(12) A steel frame shown in Fig. 2.23 above supports a machine which exerts horizontal force of 45000 sin15t (N). Find steady-state amplitude. Assume a damping ratio η as 0.09. Take EI = 31.5 × 10⁶ N-m².

Solution

$$k = 2 \times \frac{3EI}{h^3} = \frac{3 \times 2 \times 31.5 \times 10^6}{4.5^3} = 2.07 \times 10^6 \ N/m$$

Natural frequency

$$\omega = \sqrt{\frac{k}{m}} = \sqrt{\frac{2.07 \times 10^6 \times 9.81}{45000}} = 21.24 \ rad/sec$$

Frequency ratio r is given by

$$r = \frac{\Omega}{\omega} = \frac{15}{21.24} = 0.706$$

substituting all the values in equation, the steady-state amplitude is

$$\overline{X} = \frac{F/k}{\sqrt{(1-r^2)^2 + (2\eta r)^2}} = \frac{45000}{2.07 \times 10^6 \times \sqrt{(1-0.706^2)^2 + (2 \times 0.09 \times 0.706)^2}}$$

$$\overline{X} = 0.042 \ m = 42.01 \ mm$$

2.8 EXERCISE PROBLEMS

(1) A mass is connected by three linear springs as shown in Fig. 2.24. Determine the natural frequency and time period.

(2) Determine the natural frequency and natural period of the system consisting of a mass of 100 kg attached to a horizontal cantilever beam through the linear string. The cantilever beam has a thickness of 8 mm and width of 120 mm. E = 2.1 × 10⁵ N/mm², L = 700 mm, k = 10 N/mm.

Single Degree of Freedom Systems

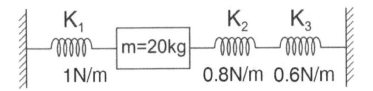

FIGURE 2.24: Spring Mass system

(3) Determine the natural period of the system shown in Fig. 2.25. Assume beam and spring supporting are massless. EI = constant.

(4) If in Problem 3, L = 2.5 m, W = 2500 N, EI = 11.37 × 10^6 N-mm², k_{spring} = 350 kN/m. Determine the displacement and velocity after 1 sec. Take initial displacement x_0 = 25.4 mm and initial velocity v_0 = 510 mm/sec.

(5) Determine the natural frequency for the horizontal motion of the steel frame shown in Fig. 2.26. Assume horizontal girder to be infinitely rigid and neglect mass of columns. Take EI = 11.36 × 10^6 N-m².

(6) A spring mass system has a stiffness of 'k' N/m and a mass of 'm' kg. The frequency of system is 15 Hz. If an extra 15 kg of mass is coupled to 'm', the frequency reduces by 3 Hz. Find 'k' and 'm'.

(7) A vibrating system is designed with the following parameters, k = 160 N/m, c = 5 N - S/m, m = 2 kg, initial amplitude of vibration is 0.5. Find the decrease in amplitude from initial value after 1 cycle of vibration and the frequency of vibration.

(8) A vibrating system consisting of weight W = 500 N and a spring of stiffness k = 15 N/mm is viscously damped so that the ratio of consecutive amplitude is 1 to 0.85. Determine (a) Natural frequency of undamped system, (b) logarithmic decrement, (c) damping ratio, (d) damping coefficient.

(9) Damped vibration of a system shows the following data:

Amplitude after 2nd cycle x_2 = 10 mm
Amplitude after 3rd cycle x_3 = 7.5 mm

Take spring constant k = 7840 N/m, mass m = 7 kg, find damping coefficient c.

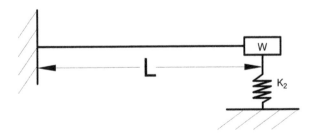

FIGURE 2.25: Spring supported cantilever beam

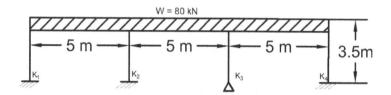

FIGURE 2.26: Steel frame

(10) A SDF system consisting of 40 kg, a spring of stiffness 2200 N/m and a dashpot with damping coefficient of 50 N-sec/m is subjected to a harmonic excitation of F = 200 Sin 5t. Write the complete solution of the equation of motion.

3 Two Degree of Freedom Systems

Earlier, we understood that a single-storey building frame can be represented by an equivalent single degree of freedom spring mass system (Den Hartog, 1985). Similarly, a two-storey frame can be represented by a two degree of freedom (2DOF) system, a three-storey frame can be by a three degree of freedom system and so on and an n-storey frame can be represented by an N degree of freedom system provided that the assumptions made in the SDF system are made again here, i.e., the storey masses are lumped and the rotations of the columns are restrained at the floor and roof levels. We can include or not include the dampers in the spring mass system depending on choice. However, damping has to be included to be realistic. A damping of about $\eta = 5\%$ is generally assumed for concrete frames, a little lower value of $\eta = 2\%$ for steel. Again, in steel frames, the damping depends on whether the joints are bolted, riveted or welded.

In bolted and riveted joints, and especially in bolted joints, friction between the nuts and the surface adds to the damping. Now let us see how we model a two-storey frame. Let us assume that the damping is neglected (Fig. 3.1).

In any dynamic analysis, the first step is to identify the degrees of freedom and then to write the equations of motion. There will be as many equations of motion as the number of degrees of freedom. Therefore, there are two equations of motion in the present case.

The equations of motion are

$$m_1\ddot{x}_1 + (k_1+k_2)x_1 - k_2x_2 = 0 \tag{3.1}$$

$$m_2\ddot{x}_2 - k_2x_1 + k_2x_2 = 0 \tag{3.2}$$

Then the next step is to solve the free vibration problem to obtain the characteristics of the systems like the natural frequencies and mode shapes.

To obtain the free vibrations solution, as before, we assume harmonic motion of the masses which always is the fact when it is freely vibrating. Thus, we assume,

$$x_1(t) = \bar{x}_1 f(t) \tag{3.3}$$

$$x_2(t) = \bar{x}_2 f(t) \tag{3.4}$$

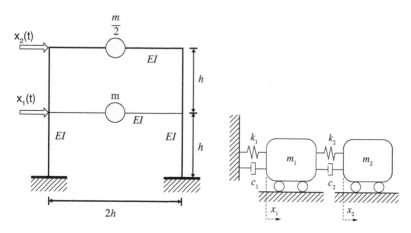

FIGURE 3.1: Two-storey frame 2DOF system

Where $\bar{x}_1$ and $\bar{x}_2$ are amplitudes which are not time dependent and are time functions. We may observe that it is the same for both x_1 and x_2 because in any particular mode, both the masses vibrate synchronously as the motion is harmonic.

$$m_1\bar{x}_1\ddot{f}(t) + (k_{11}\bar{x}_1 + k_{12}\bar{x}_2)f(t) = 0 \tag{3.5}$$

$$m_2\bar{x}_2\ddot{f}(t) + (k_{12}\bar{x}_1 + k_{22}\bar{x}_2)f(t) = 0 \tag{3.6}$$

Here, $k_{11} = k_1 + k_2, k_{22} = k_2, k_2 = -k_{12} = -k_{21}$.
The above notations are introduced to familiarize ourselves with matrix notations.
From Equations 3.5 and 3.6,

$$-\frac{\ddot{f}(t)}{f(t)} = \frac{k_{11}\bar{x}_1 + k_{12}\bar{x}_2}{m_1\bar{x}_1} = \frac{k_{12}\bar{x}_1 + k_{22}\bar{x}_2}{m_2\bar{x}_2} = \beta \tag{3.7}$$

where β is a constant, which is quite natural because all the quantities like $k_{11}, k_{12}, \bar{x}_1, \bar{x}_2$ are constants.

$$\therefore \ddot{f}(t) + \beta f(t) = 0 \tag{3.8}$$

Also,

$$(k_{11} - \beta m_1)\bar{x}_1 + k_{12}\bar{x}_2 = 0 \tag{3.9}$$

$$k_{12}\bar{x}_1 + (k_{22} - \beta m_2)\bar{x}_2 = 0 \tag{3.10}$$

substituting, $f(t) = Be^{\alpha t}$ in Equation 3.8

$$\alpha^2 + \beta = 0 \tag{3.11}$$

Two Degree of Freedom Systems

$$\alpha_{1,2} = \pm\sqrt{-\beta} \tag{3.12}$$

$$f(t) = B_1 e^{\alpha_1 t} + B_2 e^{\alpha_2 t} \tag{3.13}$$

$$f(t) = B_1 e^{\sqrt{-\beta}t} + B_2 e^{-\sqrt{-\beta}t} \tag{3.14}$$

β has to be necessarily positive, or otherwise we will not be able to realize oscillatory motion. Therefore, presuming that β as positive, letting $\beta = \omega^2$, where ω is another constant, $\alpha_{1,2} = \pm i\omega$, where $i = \sqrt{-1}$

$$f(t) = B_1 e^{i\omega t} + B_2 e^{-i\omega t} \tag{3.15}$$

Here we know that $e^{\pm i\omega t} = cos\omega t \pm i sin\omega t$

$$f(t) = (B_1 + B_2)cos\omega t + i(B_1 - B_2)sin\omega t \tag{3.16}$$

Letting $(B_1 + B_2) = Dcos\theta, i(B_1 - B_2) = Dsin\theta$

$$f(t) = D(cos\omega t - \theta) \tag{3.17}$$

Again, D is another constant. We say that ω is the frequency of the harmonic motion; θ is its phase angle. They are the same for both the masses or for both x_1 and x_2.

In the equations 3.9 and 3.10, substitute ω^2 for β. Thus we obtain the following equations 3.18 and 3.19.

$$(k_{11} - \omega^2 m_1)\bar{x}_1 + k_{12}\bar{x}_2 = 0 \tag{3.18}$$

$$k_{12}\bar{x}_1 + (k_{22} - \omega^2 m_2)\bar{x}_2 = 0 \tag{3.19}$$

The above are two simultaneous algebraic equations. They form what is called the well-known eigenvalue problem. The task before us is to find the value of ω^2 for a non-trivial solution of the above equation, which means that we find the values of the natural frequencies of the system and here there are two values ω_1^2 and ω_2^2, the frequencies of the two modes, the lowest ω_1 is called the fundamental natural frequency of the first mode and the next higher one ω_2 is the natural frequency of the second mode. After finding ω_1^2 and ω_2^2 we find the values of $\bar{x}_1, \bar{x}_2$ for each of the values of ω, i.e., ω_1 and ω_2 from Equations 3.18 and 3.19, which means that we have determined $\bar{x}_1$ and $\bar{x}_2$ in two modes, mode 1 and mode 2, respectively. But here we cannot obtain absolute values, instead, we find the ratios $\bar{x}_1/\bar{x}_2$. We require only that much, as we can easily plot the mode shapes which are unique by knowing the ratios. We cannot obtain absolute values because the right-hand sides of the Equations 3.18 and 3.19 are zero or in other words it is a free vibration problem.

For the non-trivial solution, we require,

$$\Delta(\omega^2) det \begin{bmatrix} k_{11} - \omega^2 m_1 & k_{12} \\ k_{12} & k_{22} - \omega^2 m_2 \end{bmatrix} = 0 \tag{3.20}$$

The above determinant is called the characteristic determinant. Expanding the determinant, we get an equation called the characteristic equation. It is called charactersic because it gives the chararterstics of the system which are the eigenvalues (natural frequencies) and eigenvectors (mode shapes). The latter depend on the properties viz. mass and stiffness of the system and obviously do not vary either with force or with time. It is a quadratic or second degree in ω^2 and therefore we get two values of ω^2. Here it is quadratic because it is a two degree of freedom system. In a three degree of freedom system, it will be a cubic, equation and in a N degree of system it will be a Nth degree polynomial equation yielding N eigenvalues and eigenvectors. The equation is

$$m_1 m_2 (\omega^2)^2 - (m_1 k_{22} + m_1 k_{11})\omega^2 + k_{11} k_{22} - k_{12}^2 = 0 \tag{3.21}$$

$$\omega_{1,2}^2 = \frac{1}{2}\left(\frac{m_1 k_{22} + m_2 k_{11}}{m_1 m_2}\right) \mp \frac{1}{2}\sqrt{\left(\frac{m_1 k_{22} + m_2 k_{11}}{m_1 m_2}\right)^2 - 4\left(\frac{k_{11} k_{22} - k_{12}^2}{m_1 m_2}\right)} \tag{3.22}$$

Having determined ω_1^2 and ω_2^2, or in other words, ω_1 and ω_2, we now find the values of $\bar{x}_1/\bar{x}_2$. The equations are homogeneous and therefore we cannot have absolute values for $\bar{x}_1$ and $\bar{x}_1$. We will have only relative values. Therefore, we find the ratios of $\bar{x}_1/\bar{x}_2$ for each of the values of ω, viz., ω_1 and ω_2. The ratio for ω_1 is for the first mode and for the second mode, the ratio is found out for ω_2.

$$\left(\frac{\bar{x}_2}{\bar{x}_1}\right)_{1^{st} mode} = -\frac{k_{11} - \omega_1^2 m_1}{k_{12}} = \frac{k_{12}}{k_{22} - \omega_1^2 m_2} \tag{3.23}$$

$$\left(\frac{\bar{x}_2}{\bar{x}_1}\right)_{2^{st} mode} = -\frac{k_{11} - \omega_2^2 m_1}{k_{12}} = \frac{k_{12}}{k_{22} - \omega_2^2 m_2} \tag{3.24}$$

The $\left(\frac{\bar{x}_1}{\bar{x}_2}\right)_{1^{st} mode}$ and $\left(\frac{\bar{x}_1}{\bar{x}_2}\right)_{2^{st} mode}$ are called eigenvectors in the 1st and 2nd mode, respectively. Their plots show the mode shapes.

3.1 FORCED RESPONSE OF DAMPED TWO-STOREY BUILDINGS

Equations of motion (Fig. 3.2).

$$m_1 \ddot{x}_1 + (c_1 + c_2)\dot{x}_1 + (k_1 + k_2)x_1 - c_2 \dot{x}_2 - k_2 x_2 = \bar{F}_1 \sin\Omega(t) \tag{3.25}$$

$$m_2 \ddot{x}_2 - c_2 \dot{x}_1 - k_2 x_1 + c_2 \dot{x}_2 + k_2 x_2 = \bar{F}_2 \sin\Omega(t) \tag{3.26}$$

The solution in general can be expressed in the form of

$$x_1 = \bar{x}_1 e^{i\Omega t} \quad (a) \quad and \quad x_2 = \bar{x}_2 e^{i\Omega t} \quad (b)$$

Two Degree of Freedom Systems

differentiating (a) and (b) w.r.t. 't' and substituting in 3.25 and 3.26

$$\dot{x}_1 = i\Omega \bar{x}_1 e^{i\Omega t}, \quad \text{and} \quad \ddot{x}_1 = -\Omega^2 \bar{x}_1 e^{i\Omega t} \qquad (3.27)$$

$$\dot{x}_2 = i\Omega \bar{x}_2 e^{i\Omega t}, \quad \text{and} \quad \ddot{x}_2 = -\Omega^2 \bar{x}_2 e^{i\Omega t} \qquad (3.28)$$

$$\text{knowing that}, e^{\pm i\theta} = \cos\theta \pm i\sin\theta, \text{where} \quad \theta = \Omega t \qquad (3.29)$$

$$\Rightarrow m_1(-\Omega^2 \bar{x}_1 e^{i\Omega t}) + (c_1 + c_2)(i\Omega \bar{x}_1 e^{i\Omega t}) + (k_1 + k_2)\bar{x}_1 e^{i\Omega t} - c_2(i\Omega \bar{x}_2 e^{i\Omega t}) - k_2(\bar{x}_2 e^{i\Omega t}) = \overline{F}_1 e^{i\Omega t}$$

$$\Rightarrow (k_1 + k_2 - m_1\Omega^2)\bar{x}_1 + (c_1 + c_2)(i\Omega \bar{x}_1) - (k_2 + c_2 i\Omega)\bar{x}_2 = \overline{F}_1 \qquad (3.30)$$

Similarly,
$$\Rightarrow m_2(-\Omega^2 \bar{x}_1 e^{i\Omega t}) - c_2(i\Omega \bar{x}_1 e^{i\Omega t}) - k_2(\bar{x}_1 e^{i\Omega t}) + c_2(i\Omega \bar{x}_2 e^{i\Omega t}) + k_2(\bar{x}_2 e^{i\Omega t}) = \overline{F}_2 e^{i\Omega t}$$

$$\Rightarrow -[k_2 + c_2(i\Omega)]\bar{x}_1 + [c_2 i\Omega + k_2 - m_2\Omega^2]\bar{x}_2 = \overline{F}_2 \qquad (3.31)$$

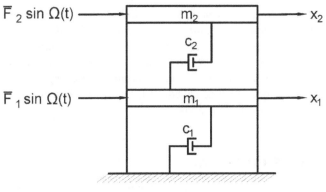

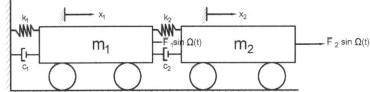

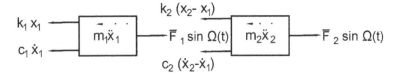

FIGURE 3.2: Two storey frame, two degree of freedom model and free body forces

Solving the above Equations 3.30 and 3.31, we obtain the solution for $\bar{x}_1$ and $\bar{x}_2$ in complex forms. Later we separate them in two parts, real and imaginary, and keep them separate. In a problem, if the forcing function contains cos Ωt we will retain the real part, while if the forcing function contains sin Ωt we will retain the imaginary component. Another interesting thing we may note is that we are able to get the solution very easily because we have substituted the unknown response in the form of $\bar{x}e^{i\Omega t}$ because we know that the time function of the response will be the same as that of the force in steady state. In other words, the frequency of the response in steady state will be Ω which is the same as forcing frequency. In short, during steady state the system vibrates with the frequency of the force. It goes without mention that the vibrations due to initial conditions x_0 and $\dot{x}_0$ would have decayed and died due to damping.

(As the author had earlier said that he would always like to see structural dynamics in real life and give examples accordingly in the above case also he would give the following example. All of us are born with certain initial conditions on account of our place of birth and parentage. Till we go to initial schooling, we will vibrate only with our habits picked up in our childhood. As we grow further, we pick up vibrations from the environment around us and during our growth till adulthood we will be vibrating with a combination of our habits due to our place of birth and parentage and also those due to the environment around us. This is like a system vibrating both due to initial conditions and force. Much later in adulthood mostly habits picked up from the environment will stay and very little or zero will from our childhood. It depends on the damping ratio in the person. It is like only steady state.) Later we see that if we have to get a general solution for any kind of arbitrarily loading which may include sinusoidal also we adopt a method called modal analysis, where we need not assume any known time function as done above.

3.2 EXERCISE PROBLEMS

(1) For the frame shown in Fig. 3.1, take column dimensions as 400 mm × 400 mm, beam as 300 mm × 300 mm and slab thickness as 125 mm. Find, natural frequencies and periods. Take live load as 3 kN/m² and concrete and steel grades as M 25 and Fe 415, respectively. Take bay width as 3m and storey height as 3 m.

(2) For the same frame as in Example 1, consider three storeys and three bays and find natural frequencies and mode shapes. Take bay width as 4 m and storey height as 3 m.

(3) Find the steady-state motion of a two-storey frame as shown in Figure 3.1 for the following load. $F_1 = 45 \sin 5t$ and $F_2 = 55 \sin 5t$.

(4) Find the natural frequencies and mode shapes of a four-storey frame with 3 bay having bay width of 4 m and a floor height of 3.5 m. Take live load as 2 kN/m² and floor finish as 1.5 kN/m², and terrace live load as 1.5 kN/m² and floor finish as 3 kN/m². Assume column dimensions as 300 mm × 450 mm, beam as 300 mm × 300 mm and slab thickness as 150 mm. Take concrete and steel grades as M 30 and Fe 415, respectively.

4 Force Transmitted to the Support

We have seen that the force transmitted to the foundation with a flexible spring between the main system and the foundation will be relatively low compared to the case when the spring is relatively rigid. In our daily life, a similar situation is encountered. If there is a hard interface between two persons A and B in the form of a person C who is very rigid in his approach, then whatever transfers between A and C will be straight away transmitted to B. However, if C is very flexible and can absorb a lot of hardships from A, C does not reveal to B, but rather absorbs.

Yet another useful derivation is about the force transmitted to the supports. The concept leads to vibration isolation. We have seen just now the response of a single degree of freedom system subjected to support motion. There, the support motion was expressed in terms of support acceleration. It was found that the support acceleration was finally equivalent to an external force on the mass. Now instead, the support motion can also be expressed in terms of support velocity and displacement. Then, the equation of motion is expressed in terms of absolute displacement, velocity and acceleration. Support displacement and velocity will be on the R.H.S of the equation representing the force.

$$m\ddot{x} + c(\dot{x} - \dot{x}_s) + k(x - x_s) = 0 \quad (4.1)$$

$$m\ddot{x} + c\dot{x} + kx = c\dot{x}_s + kx_s \quad (4.2)$$

Let the support displacement be harmonic, $x_s(t) = \bar{x}_s \sin\Omega t$, $\bar{x}_s =$ amplitude of support displacement and Ω is the forcing frequency.

Substituting in Equation (4.2), we get

$$m\ddot{x} + c\dot{x} + kx = c\bar{x}_s \Omega \cos\Omega t + k\bar{x}_s \sin\Omega t$$

$$= \bar{F} \sin(\Omega t + \alpha) \quad (4.3)$$

Letting,

$$\bar{F} \sin\alpha = c\bar{x}_s \Omega \quad \text{and} \quad \bar{F} \cos\alpha = k\bar{x}_s$$

We get,

$$\bar{F} = \sqrt{(c\bar{x}_s \Omega^2) + (k\bar{x}_s)^2}$$

$$F = \bar{x}_s k\sqrt{1+(2r\eta)^2}$$

where, $r = \Omega/\omega, \eta = c/c_{cr}$ $tan\alpha = \frac{c\Omega}{k} = 2r\eta$

$$\frac{F}{k} = \bar{x}_s\sqrt{1+(2r\eta)^2} \tag{4.4}$$

The above Equation (4.3) is also in the form of an equation for an oscillator subjected to harmonic force on the mass.
The steady-state solution

$$x(t) = \frac{(\bar{F}/k)sin(\Omega t + \alpha - \theta)}{\sqrt{(1-r^2)^2+(2r\eta)^2}} \tag{4.5}$$

Using Equation (4.4).

$$\frac{x(t)}{\bar{x}_s} = \frac{\sqrt{1+(2r\eta)^2}}{\sqrt{(1-r^2)^2+(2r\eta)^2}} sin(\Omega t + \alpha - \theta) \tag{4.6}$$

$$\frac{x(t)}{\bar{x}_s} = \frac{\bar{x}}{\bar{x}_s} sin(\Omega t + \alpha - \theta) \tag{4.7}$$

where, $\bar{x}$ = amplitude of steady-state oscillation, $\bar{x}_s$ = amplitude of harmonic support displacement.

$$T_r = \frac{\bar{x}}{\bar{x}_s} = \frac{\sqrt{1+(2r\eta)^2}}{\sqrt{(1-r^2)^2+(2r\eta)^2}} \tag{4.8}$$

where T_r is called the transmissibility gives the ratio of amplitude of oscillator mass displacement to that of support oscillation. It is required to design the oscillator spring and mass in such a way that it is isolated from the support disturbances. It is called vibration isolation. The plot of T_r vs r is shown in Fig. 4.1. All the curves pass through a point where $r = \sqrt{2}$.

The force transmitted to the foundation.

$$F_t = kx + c\dot{x} \tag{4.9}$$

Substituting for x and $\dot{x}$, where

$$x = \bar{x}sin(\Omega t - \theta) \tag{4.10}$$

$$F_t = \bar{x}\sqrt{k^2+c^2\Omega^2} sin(\Omega t + \alpha - \theta) \tag{4.11}$$

$$tan\alpha = \frac{c\Omega}{k} = 2\eta r \tag{4.12}$$

Force Transmitted to the Support

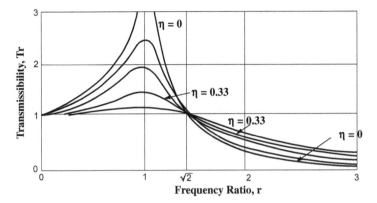

FIGURE 4.1: Transmissibility vs. frequency ratio

F_t is the expression of the harmonic forced oscillator of the support. The amplitude of the above harmonic forced oscillation is

$$\overline{F}_t = \overline{F} \frac{\sqrt{1+(2r\eta)^2}}{\sqrt{(1-r^2)^2+(2r\eta)^2}} \quad (4.13)$$

$$T_r = \frac{\overline{F}_t}{\overline{F}} = \frac{\sqrt{1+(2r\eta)^2}}{\sqrt{(1-r^2)^2+(2r\eta)^2}} \quad (4.14)$$

It is better to recall that the force acting on the mass is $\overline{F}sin\Omega t$ and

$$\overline{x} = \frac{\overline{F}/k}{\sqrt{(1-r^2)^2+(2r\eta)^2}}, \quad x = \overline{x}sin(\Omega t - \theta) \quad (4.15)$$

$$tan\theta = \frac{2\eta r}{1-r^2} \quad (4.16)$$

It may be surprising to see that the same expression represents both the transmissibilities, the one from the support to the oscillator and the other from the oscillator to the support. The concept is very useful in vibration oscillation, the former in the system subjected to earthquake ground motion and the latter in the machine foundations to relieve the forces on the foundation.

Once again I would like to cite a nice experience most of us would have gone through while we were young. When we were asked to fetch milk in a metal container with a handle on a bicycle if we had suspended the container from the handle of the cycle there would not be any milk left by the time we came home. It was because the vibrations of the handle due roughness of the road would be transferred to the container completely. We would have sensed it immediately and instead of suspending the handle to the handle of the cycle we carried it by one hand, riding the cycle only by the other hand we would have saved most of the milk. That was because we kept adjusting the stiffness and damping of the hand holding the container by bending the hand to the required levels. We did that by sheer fuzzy logic or sixth sense as we call. What exactly we did then was vibration isolation. The force transmitted to the container was controlled by stiffness and damping of the support which was our hand holding the container. Here it was a case of support accelerations created by the roughness of the road through the cycle wheel tyres and suspensions. Of course we could have controlled by reducing the speed of the cycle, by proper inflation of the tyres and of course by riding on a smoother road.

4.1 EXERCISE PROBLEMS

(1) A motor rotating at a frequency of 75 rpm, weighing 35 kN is mounted on a simply supported beam of span 5 m. Take damping as 5% and find steady-state response and force transmitted to support. Also find phase angle.

(2) Consider a two-storeys frame having columns 400 mm × 400 mm and beams as 300 mm × 300 mm and slab thickness as 125 mm. If a motor is placed on the 2nd floor having frequency of 50 rad/sec and weight of 30 kN, find the force transmitted to the base. Take height of storey as 3 m and span of beams as 3 m.

(3) Take ground storey height as 4.5 m in Example 2 and find the force transmitted to the foundation if a motor kept on the top floor rotates at 300 rpm. Take damping ratio η as 0.08. *Note*. In all the problems above take an unbalanced weight of 5 kN at an eccentricity of 0.2 m.

5 Duhamel's Integral

We have become quite familiar with harmonic loading in the earlier chapters (Biggs, J M 1964). Harmonic loadings expressed as $\overline{F}sin\Omega t$ or $\overline{F}cos\Omega t$ or both $\overline{F}(sin\Omega t + cos\Omega t)$ or $\overline{F}_1 sin\Omega t + \overline{F}_2 cos\Omega t$ are observed in our daily life. The loading generated by centrifugal and reciprocating pumps, vibrational force induced by vortex shedding due to wind encountering a bluff body, and loads due to machines on foundation are all examples of harmonic loading. Earthquake loading could be considered as random harmonic loading. But we will not be dealing with such loads always. We sometimes come across suddenly applied loads held constant, impulsive loads, which are very large loads applied over very small duration. Pulse loads having load-time functions of different shapes like rectangular, triangular and also loads which gradually increase and remain constant called ramp functions and so on, or loads having time functions of any other arbitrary shape are also possible. In all such cases, we employ another technique called "Duhamel's Integral" to obtain the response. Nevertheless, the integral can also be elegantly employed in the case of harmonic and earthquake ground motions.

To describe the integral, we need to start with the concept of impulse loading. Shown in Fig. 5.1 is an impulse of the force, F(t), applied on a spring mass system over a very small duration, dt, denoted by F(t)dt.

The impulse is applied so suddenly that there will be hardly any time for the spring to respond and therefore the initial displacement due to force application is assumed to be zero, while there is only initial velocity $\dot{x}(0) = \dot{x}_0$. (The author always liked to give examples from day-to-day life in trying to drive home the physics of the problem. For example, here it is like a person saying that he hardly had any time to react when he was taken by surprise.) From Newton's Law of Motion,

$$m\frac{d}{dt}(\dot{x}) = F(t) \qquad (5.1)$$

Hereafter, for convenience, use a dummy time variable 'τ' and let us keep symbol 't' for the integration limits.

$$\therefore m\frac{d}{d\tau}(\dot{x}) = F(\tau) \qquad (5.2)$$

$$d(\dot{x}) = \frac{F(\tau)d\tau}{m} \qquad (5.3)$$

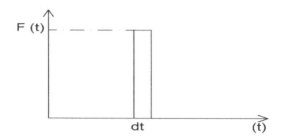

FIGURE 5.1: Impulse loading

$F(\tau)\, d(\tau)$ is the incremental impulse at any time τ, which causes an incremental velocity $d(\dot{x})$. The incremental velocity $d(\dot{x})$ could be considered as initial velocity due to impulse applied at any time τ in Fig. 5.2.

In the entire time history of the arbitrary loading shown in Fig. 5.2, there are an infinite number of strips shown as shaded. The response of undamped oscillator due to the impulse $F(\tau)\, d\tau$ will be obtained in terms of the initial velocity and displacement at $t = \tau$, which are

$$d\dot{x} = \dot{x}_\tau = \frac{F(\tau)d\tau}{m} \qquad (5.4)$$

and $x_\tau = 0$. ∴ We know that the response of an undamped oscillator with initial condition x_0 and $\dot{x}_0$ is

$$x(t) = x_0 \cos\omega t + \frac{\dot{x}_0}{\omega}\sin\omega t \qquad (5.5)$$

as observed above, the system has hardly any time to react and therefore the displacement of the spring is zero. Therefore $x_0 = x_\tau = 0, \dot{x}_0 = \dot{x}_\tau$.

$$x(t) = \frac{\dot{x}_0}{\omega}\sin\omega t = \frac{\dot{x}_\tau}{\omega}\sin\omega t \qquad (5.6)$$

$$= \frac{F(\tau)d\tau}{m\omega}\sin\omega(t-\tau) \qquad (5.7)$$

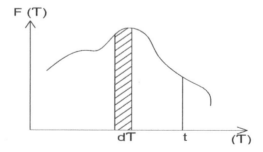

FIGURE 5.2: Arbitrary loading

Duhamel's Integral

It is $(t - \tau)$, because the system responding after the application of impulse at $t = \tau$. There are an infinite number of strips and therefore the total response is

$$x(t) = \frac{1}{\omega} \int_0^t \frac{F(\tau)d\tau}{m} \sin\omega(t - \tau) = \frac{1}{m\omega} \int_0^t F(\tau)\sin\omega(t - \tau)d\tau \qquad (5.8)$$

The above is due to only loading. In addition, there may be initial conditions x_0 and $\dot{x}_0$ and the total response is

$$x(t) = x_0 \cos\omega t + \frac{\dot{x}_0}{\omega} \sin\omega t + \frac{1}{m\omega} \int_0^t F(\tau)\sin\omega(t - \tau)d\tau \qquad (5.9)$$

The above Equation 5.9 is called Duhamel's Integral.

Different types of dynamic loads (Clarence W de Silva, 2005, Bangash M Y H 2008).

In all dynamic problems we require to deal with, at first, the applied force vs. time curve. This could be of several forms as shown below (Figs. 5.3 through 5.7):

In all the above cases, Duhamel's Integral is an elegant technique to obtain the solution.

Example 1. A suddenly applied load is acting on an undamped SDF system. Find the response. (Fig. 5.3 shows the suddenly applied load held constant and Fig. 5.8 shows the SDF system on which the load is applied).

Solution:

The equation of motion: $m\ddot{x} + kx = F(t)$
$F(t) = \overline{F}$ at $t = 0$
Let the initial conditions be 'x_0' and '$\dot{x}_0$' at $t = 0$
The general solution is

$$x(t) = x_0 \cos\omega t + \frac{\dot{x}_0}{\omega} \sin\omega t + \frac{1}{m\omega} \int_0^t F(\tau)\sin\omega(t - \tau)d\tau \qquad (5.10)$$

the particular integral is

$$x(t) = \frac{1}{m\omega} \int_0^t F(\tau)\sin\omega(t - \tau)d\tau \qquad (5.11)$$

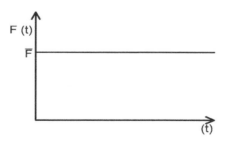

FIGURE 5.3: Load suddenly applied and held constant is $\overline{F}$ = Max. load

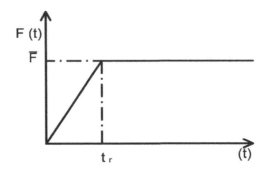

FIGURE 5.4: Gradually applied and held constant (Ramp Form) t_r = rise time

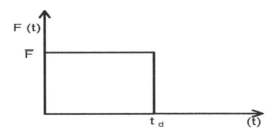

FIGURE 5.5: Rectangular pulse t_d = duration

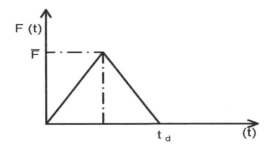

FIGURE 5.6: Triangular pulse t_d = duration

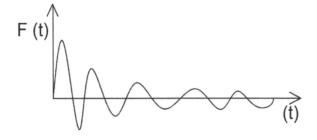

FIGURE 5.7: Earthquake loading

Duhamel's Integral

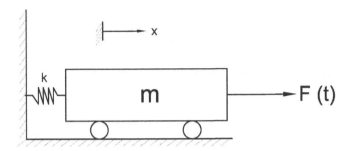

FIGURE 5.8: Undamped SDF system

$$= \frac{1}{m\omega} \int_0^t \overline{F} \sin\omega(t-\tau) d\tau \qquad (5.12)$$

$$= \frac{\overline{F}}{m\omega^2} [\cos\omega(t-\tau)]_0^t = \frac{\overline{F}}{m\omega^2}(1-\cos\omega t) \qquad (5.13)$$

Total solution

$$x(t) = x_0 \cos\omega t + \frac{\dot{x}_0}{\omega} \sin\omega t + \frac{\overline{F}}{k}(1-\cos\omega t) \qquad (5.14)$$

when, ' x_0 ' = ' $\dot{x}_0$ '= 0

$$x(t) = \frac{\overline{F}}{k}(1-\cos\omega t) \qquad (5.15)$$

Alternatively,
Let us approach by classical solution

$$m\ddot{x} + kx = \overline{F} \qquad (5.16)$$

particular integral is (by inspection) $x = \overline{F}/k$ ∴ general solution.

$$x = A\cos\omega t + B\sin\omega t + \frac{\overline{F}}{k} \qquad (5.17)$$

$$\begin{bmatrix} x = x_0 \\ \dot{x} = \dot{x}_0 \end{bmatrix} At \ t = 0 \qquad (5.18)$$

$$x_0 = A + \frac{\overline{F}}{k} \Rightarrow A = x_0 - \frac{\overline{F}}{k} \qquad (5.19)$$

And,

$$\dot{x}_0 = B\omega \Rightarrow B = \frac{\dot{x}_0}{\omega} \tag{5.20}$$

If $x_0 = \dot{x}_0 = 0$, then $A = -(\overline{F}/k)$ and $B = 0$. Thus,

$$x = \frac{\overline{F}}{k}(1 - \cos\omega t) \tag{5.21}$$

$$\frac{\overline{F}}{k} = x_{st} \tag{5.22}$$

Triangular Pulse (Blast Load) (See Fig. 5.9)

$$\frac{x}{x_{st}} = 1 - \cos\omega t \tag{5.23}$$

When there is a distant explosion, the pressure waves travel in all the directions, and when they encounter an obstacle the waves impinge on the obstacle. For example a building experiences a pressure higher than the atmospheric pressure called "overpressure." In order to analyze such a building for the blast effects, overpressure which depends on the size of the explosion and distance from the building caused by the blast, will be applied. The overpressure vs. time plots will be assumed to be in the form of the triangle as shown in Fig. 5.10.

Both $\overline{F}$ and t_d depend on the size of the explosion and distance from the building.

Example 2. Response of undamped SDF to triangular pulse (Fig. 5.10).

The equation of the straight-line sloping from a value $\overline{F}$ to zero over t_d is

$$F(\tau) = \overline{F}(1 - \frac{\tau}{t_d}) \tag{5.24}$$

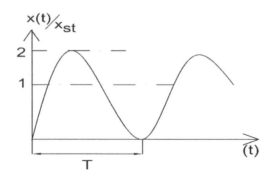

FIGURE 5.9: x_{st} = static deflection, T = time period

Duhamel's Integral

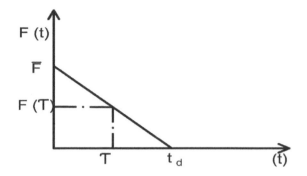

FIGURE 5.10: $\overline{F}$ is the maximum overpressure, t_d = duration

Let $x_0 = \dot{x}_0 = 0$

$$x(t) = \frac{1}{m\omega} \int_0^t F(\tau) \sin\omega(t-\tau) d\tau \qquad (5.25)$$

$$x(t) = \frac{\overline{F}}{m\omega} \int_0^t (1 - \frac{\tau}{t_d}) \sin\omega(t-\tau) d\tau \qquad (5.26)$$

$$= \frac{\overline{F}}{m\omega} [\frac{\cos\omega(t-\tau)}{\omega} - \frac{\tau}{t_d}\frac{\cos\omega(t-\tau)}{\omega} + \frac{1}{t_d}\int \frac{\cos\omega(t-\tau)}{\omega} dt]_0^t \qquad (5.27)$$

$$= \frac{\overline{F}}{m\omega} [\frac{\cos\omega(t-\tau)}{\omega} - \frac{\tau}{t_d}\frac{\cos\omega(t-\tau)}{\omega} + \frac{1}{t_d}\frac{\sin\omega(t-\tau)}{\omega^2}]_0^t \qquad (5.28)$$

$$= \frac{\overline{F}}{m\omega^2}[1 - \cos\omega t - \frac{t}{t_d} + \frac{\sin\omega t}{\omega t_d}] \qquad (5.29)$$

$$= [\frac{\overline{F}}{k}(1 - \cos\omega t) + \frac{\overline{F}}{kt_d}(\frac{\sin\omega t}{\omega} - t)] \qquad (5.30)$$

Now we substitute $t = t_d$ in the above equation as well as in $\dot{x}(t)$ to obtain $x_{t=t_d}$ and $\dot{x}_{t=t_d}$ as the initial condition, and the rest of the vibrations are free vibrations with initial conditions $x_{t=t_d}$ and $\dot{x}_{t=t_d}$.

The above solution as given by Equation 5.30 is valid between times 0 and t_d. The system continues to vibrate even after the removal of the force for some more time till it comes to a complete halt due to damping present. Therefore the vibrations after t_d viz. the free vibrations can be obtained from conditions at t_d. The conditions at t_d are initial conditions for free vibrations after t_d and they will be found from the Equation 5.30.

6 Modal Analysis

6.1 MULTI-DEGREE OF FREEDOM SYSTEMS SUBJECTED TO EXTERNAL DYNAMIC FORCES: MODAL ANALYSIS

Under earthquake resistant design, the modal analysis has been demonstrated through the matrix approach. Although the matrix approach is neat and crisp, it may be difficult to understand the physics behind the approach, because obviously what is neat and short generally fails to explain the mechanics in detail. Therefore, in what follows, the same will be described in the long hand approach.

Example. Shear building subjected to forced oscillation by external dynamic forces: modal analysis.

Consider a two storeyed building frame as shown in Fig. 6.1. If the forces are harmonic, the response can be easily obtained as earlier. But if the loading at each floor level is other than harmonic, like the blast load which will be triangular pulses, modal analysis is the best method.

The equations of motion:

$$m_1 \ddot{x}_1 + (k_1 + k_2)x_1 - k_2 x_2 = F_1(t) \qquad (6.1)$$

$$m_2 \ddot{x}_2 - k_2 x_1 + k_2 x_2 = F_2(t) \qquad (6.2)$$

The above set of equations are coupled. In the modal analysis or in other words mode superposition method, we decouple the equations with the help of orthogonality conditions. To make use of orthogonality conditions, we should have already obtained natural frequencies and mode shapes by free vibration analysis. In order to decouple or uncouple the equations of motion, we transfer the equations to a new set of coordinate system. It is worth recalling that coupling or decoupling a set of equations is not by virtue of the properties of system, but property which depends on choice of coordinate system.

Therefore, we now use a transformation which transforms the equations of motion to another reference coordinate system so that they can be easily un-coupled by the help of orthogonality principles.

The transformation is

$$Let, x_1(t) = \bar{x}_{11} z_1(t) + \bar{x}_{12} z_2(t) \qquad (6.3)$$

$$x_2(t) = \bar{x}_{21} z_1(t) + \bar{x}_{22} z_2(t) \qquad (6.4)$$

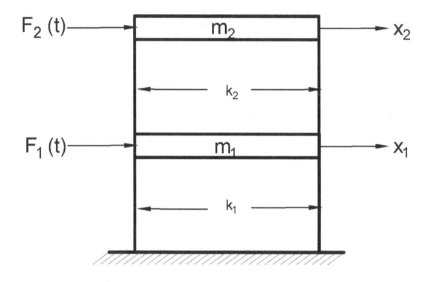

FIGURE 6.1: Two storeyed building frame

Upon substituting in Equations of motion 6.3 and 6.4

$$m_1\bar{x}_{11}\ddot{z}_1 + (k_1+k_2)\bar{x}_{11}z_1 - k_2\bar{x}_{21}z_1 + (k_1+k_2)\bar{x}_{12}z_2 - k_2\bar{x}_{22}z_2 + m_1\bar{x}_{12}\ddot{z}_2 = F_1(t)$$
(6.5)

$$m_2\bar{x}_{21}\ddot{z}_1 - k_2\bar{x}_{11}z_1 + k_2\bar{x}_{21}z_1 - k_2\bar{x}_{12}z_2 + k_2\bar{x}_{22}z_2 + m_2\bar{x}_{22}\ddot{z}_2 = F_2(t) \quad (6.6)$$

Now, we have a new set of unknowns, viz. z_1 and z_2, instead of x_1 and x_2. Of course, x_1 and x_2 can be obtained after solving for z_1 and z_2 by the relationships, (6.3 and 6.4).

The advantage now is that we have a relationship between $\bar{x}_{11}, \bar{x}_{12}, \bar{x}_{21}$ and $\bar{x}_{22}$ called the "orthogonality principle," which after proper multiplication and addition will cause coupling terms to vanish and thus leaving the equations of motion in z_1 and z_2 uncoupled.

Multiply Equation 6.5 by $\bar{x}_{11}$ and Equation 6.6 by $\bar{x}_{21}$ and add these equations. Now multiply Equation 6.5 by $\bar{x}_{12}$ and Equation 6.6 by $\bar{x}_{22}$ and add, we get Equations 6.7 and 6.8.

$$(m_1\bar{x}_{11}^2 + m_2\bar{x}_{21}^2)\ddot{z}_1 + \omega_1^2(m_1\bar{x}_{11}^2 + m_2\bar{x}_{21}^2)z_1 = \bar{x}_{11}F_1(t) + \bar{x}_{21}F_2(t) \quad (6.7)$$

$$(m_1\bar{x}_{12}^2 + m_2\bar{x}_{22}^2)\ddot{z}_2 + \omega_2^2(m_1\bar{x}_{12}^2 + m_2\bar{x}_{22}^2)z_2 = \bar{x}_{12}F_1(t) + \bar{x}_{22}F_2(t) \quad (6.8)$$

The above equations are not coupled and each of them is independent in z_2 and z_1, respectively. The terms in the brackets are constant and we are calling them as $\bar{m}_1$ and $\bar{m}_2$, respectively.

Modal Analysis

Where,

$$\overline{m}_1 = (m_1 \overline{x}_{11}^2 + m_2 \overline{x}_{21}^2)$$

$$\overline{m}_2 = (m_1 \overline{x}_{12}^2 + m_2 \overline{x}_{22}^2)$$

Where $\overline{m}_1$ and $\overline{m}_2$ are modal masses.

$$\overline{k}_1 = \omega_1^2 \overline{m}_1,$$

$$\overline{k}_2 = \omega_2^2 \overline{m}_2$$

where $\overline{k}_1$ and $\overline{k}_2$ are modal stiffnesses.

$$\overline{F}_1(t) = \overline{x}_{11} F_1(t) + \overline{x}_{21} F_2(t)$$

$$\overline{F}_2(t) = \overline{x}_{12} F_1(t) + \overline{x}_{22} F_2(t)$$

Where $\overline{F}_1$ and $\overline{F}_2$ modal forces.
We write them as,

$$\overline{m}_1 \ddot{z}_1 + \omega_1^2 \overline{m}_1 z_1 = \overline{F}_1(t) \tag{6.9}$$

$$\overline{m}_2 \ddot{z}_2 + \omega_2^2 \overline{m}_2 z_2 = \overline{F}_2(t) \tag{6.10}$$

or

$$\overline{m}_1 \ddot{z}_1 + \overline{k}_1 z_1 = \overline{F}_1(t) \tag{6.11}$$

$$\overline{m}_2 \ddot{z}_2 + \overline{k}_2 z_2 = \overline{F}_2(t) \tag{6.12}$$

where, $\overline{m}_1, \overline{m}_2, \overline{k}_1, \overline{k}_2, \overline{F}_1(t), \overline{F}_2(t)$ are modal quantities, viz. modal masses, modal stiffnesses and modal forces in respective modes and here they are mode 1 and mode 2, respectively.

Equations (6.9) and (6.10) are in mode 1 and mode 2, respectively, containing natural frequencies ω_1 and ω_1 of the respective modes. Therefore, they are not only uncoupled, but also considered as equations of single degree of freedom systems, because they contain only one frequency, viz. ω_1 or ω_2. We can easily solve like any other single degree of freedom system and we can obtain z_1 and z_2 either by classical method or by Duhamel's Integral. After obtaining z_1 and z_2, we can find from the transformation Equations 6.3 and 6.4.

Orthogonality condition between mode shapes is

$$m_1 \bar{x}_{11} \bar{x}_{12} + m_2 \bar{x}_{21} \bar{x}_{22} = 0 \tag{6.13}$$

where, modal eigenvectors

$$\begin{bmatrix} \bar{x}_{11} \\ \bar{x}_{21} \end{bmatrix} = \begin{bmatrix} Modal \ vector \ at \ mass \ 1 \\ Modal \ vector \ at \ mass \ 2 \end{bmatrix} in \ mode \ 1 \tag{6.14}$$

Similarly,

$$\begin{bmatrix} \bar{x}_{12} \\ \bar{x}_{22} \end{bmatrix} = \begin{bmatrix} Modal \ vector \ at \ mass \ 1 \\ Modal \ vector \ at \ mass \ 2 \end{bmatrix} in \ mode \ 2 \tag{6.15}$$

It should be noted that the 1st subscript denotes the mass point and the 2nd subscript denotes the mode number.

In general,

$$\sum_{i=1}^{N} (m_i \bar{x}_{in} \bar{x}_{im} = 0; for, n \neq m) \tag{6.16}$$

summation over all the masses. n and m are any two modes.

In matrix form, orthogonality conditions with respect to mass and stiffness respectively are.

$$\{\bar{x}\}_n^T [m] \{\bar{x}\}_m = 0 \tag{6.17}$$

Similarly,

$$\{\bar{x}\}_n^T [k] \{\bar{x}\}_m = 0 \tag{6.18}$$

$n \neq m$.

Modal Maxima

Very often, we are interested only in maximum values in design. For obtaining maximum values, we make use of the response spectra. Response spectra can be made available for all types of loads in standard plots of max. response vs. fundamental period. In the case of pulse loads, there will be max. response vs. ratios of either t_d/T or t_r/T, where t_d is time duration of the pulse, t_r is the rise time of a ramp function, and T is the period. In obtaining the maximum response, first we obtain $z_{1,max}(z_{1,m})$ and $z_{2,max}(z_{2,m})$ from the response spectra corresponding to T_1 and T_2 (the periods of the first two modes, respectively) and after that obtain $x_{1,max}(x_{1,m}), x_{2,max}(x_{2,m})$, from the transformation Equations 6.3 and 6.4

$$x_{1,m} = |\bar{x}_{11} z_{1,max}| + |\bar{x}_{12} z_{2,max}| \tag{6.19}$$

$$x_{2,m} = |\bar{x}_{21}z_{1,max}| + |\bar{x}_{22}z_{2,max}| \tag{6.20}$$

However, obtaining $x_{1,m}$ and $x_{2,m}$ by the above procedure is conservative and overestimates the value of $x_{1,m}$ and $x_{2,m}$. The reason is that the max. values of 1st mode and the 2nd mode do not occur at the same instant of time and therefore adding them is on the higher side. Therefore, another approach, which is well established and considered to be very probable and not as high as the former one, is called Square Root of the Sum of Squares (SRSS) which originates from the field of electrical engineering, where it is called Root Mean Square (RMS) value.
Using SRSS we get

$$x_{1,m} = \sqrt{(\bar{x}_{11}z_{1,m})^2 + (\bar{x}_{12}z_{2,m})^2} \tag{6.21}$$

$$x_{2,m} = \sqrt{(\bar{x}_{21}z_{1,m})^2 + (\bar{x}_{22}z_{2,m})^2} \tag{6.22}$$

6.2 A MULTI-STOREY BUILDING SUBJECTED TO GROUND MOTIONS: MODAL ANALYSIS

It is mandatory to design any tall building for earthquake ground motion. IS 1893-2016 is the latest revision where the specifications are contained. Certain highlights of specifications are buildings that should not be so much asymmetric in plan as to cause intolerable torsional oscillations in plan about a vertical axis. The eccentricity in plan between the mass center and the stiffness center should be kept at minimum possible. The time period of principal torsional mode should be smaller than the mode which contains the principal translational modes. Now, here we just derive equations necessary to obtain ground motion response of a multi-storey building.

Let us take a two-storey building. Let us assume that is a shear building (Fig. 6.2a and Fig. 6.2b).

Let the base motion be $x_s(t)$. The equations of motion are

$$m_1\ddot{x}_1 + k_1(x_1 - x_s) - k_2(x_2 - x_1) = 0 \tag{6.23}$$

$$m_2\ddot{x}_2 + k_2(x_2 - x_1) = 0 \tag{6.24}$$

$x_s(t)$ is the base motion

Let x_{r1} = relative displacement of first floor w.r.t. ground = $x_1 - x_s$

And, x_{r2} = relative displacement of second w.r.t. ground = $x_2 - x_s$

$$m_1(\ddot{x}_1) + k_1(x_{r1}) - k_2(x_2 - x_s + x_s - x_1) = 0 \tag{6.25}$$

$$m_1(\ddot{x}_{r1} + \ddot{x}_s) + k_1(x_{r1}) - k_2(x_{r2} - x_{r1}) = 0 \tag{6.26}$$

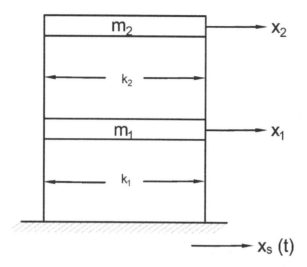

FIGURE 6.2a: Building subjected to ground motion

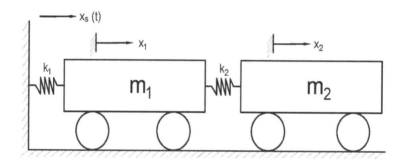

FIGURE 6.2b: Spring Mass system model

Or

$$m_1\ddot{x}_{r1} + (k_1 + k_2)x_{r1} - k_2 x_{r2} = -m_1 \ddot{x}_s \tag{6.27}$$

The second equation is

$$m_2(\ddot{x}_{r2} + \ddot{x}_s) + k_2(x_2 - x_s + x_s - x_1) = 0 \tag{6.28}$$

Or

$$m_2\ddot{x}_{r2} + k_2 x_{r2} - k_2 x_{r1} = -m_2 \ddot{x}_s \tag{6.29}$$

One interesting thing is that the right-hand sides of both of the above equations contain the same function $x_s(t)$, which means the same time function. That makes the solution very simple compared to if they were of different functions.

Modal Analysis

Using orthogonality functions as before, we uncouple the equations of motion and obtain.

$$\ddot{z}_1 + \omega_1^2 z_1 = -\frac{m_1\bar{x}_{11} + m_2\bar{x}_{21}}{m_1\bar{x}_{11}^2 + m_2\bar{x}_{21}^2}\ddot{x}_s(t) \tag{6.30}$$

$$\ddot{z}_2 + \omega_2^2 z_2 = -\frac{m_1\bar{x}_{12} + m_2\bar{x}_{22}}{m_1\bar{x}_{12}^2 + m_2\bar{x}_{22}^2}\ddot{x}_s(t) \tag{6.31}$$

Now, let us compare the RHS of Equations 6.9 and 6.10 with the RHS of Equations 6.30 and 6.31 respectively. In the former case where it was the case of a frame subjected to externally applied forces the external forces get modified after multiplication by individual eigenvectors, while in the later case the ground acceleration gets modified after multiplication by a ratio of the products of masses and eigenvectors because here the applied force is the product of mass and base acceleration. During the modal analysis, the external force here as defined above will be multiplied by a factor P_{fn} which is called a participation factor of nth mode obtained by dividing the sum of the products of various masses by the corresponding modal components of the nth mode by the sum of products of the various masses multiplied by the squares of the respective modal components of the nth mode.

P_{fn} = participation in n^{th} mode

$$\text{In } 1^{st} \text{ mode} P_{f1} = -\frac{m_1\bar{x}_{11} + m_2\bar{x}_{21}}{m_1\bar{x}_{11}^2 + m_2\bar{x}_{21}^2} \tag{6.32}$$

$$\text{In } 2^{nd} \text{ mode} P_{f2} = -\frac{m_1\bar{x}_{12} + m_2\bar{x}_{22}}{m_1\bar{x}_{12}^2 + m_2\bar{x}_{22}^2} \tag{6.33}$$

P_{fn} is called the mode participation factor. It is highest for the 1st mode and keeps reducing as mode number increases.

Let us introduce a transformation as before, to uncouple the equations of motion.

Let,

$$x_{r1} = \bar{x}_{11}z_1 + \bar{x}_{12}z_2 \tag{6.34}$$

$$x_{r2} = \bar{x}_{21}z_1 + \bar{x}_{22}z_2 \tag{6.35}$$

Let us introduce for convenience another transformation.

$$z_1 = P_{f1}q_1 \tag{6.36}$$

$$z_2 = P_{f2}q_2 \tag{6.37}$$

After substituting

$$\ddot{q}_1 + \omega_1^2 q_1 = \ddot{x}_s(t) \tag{6.38}$$

$$\ddot{q}_2 + \omega_2^2 q_2 = \ddot{x}_s(t) \tag{6.39}$$

Finally,

$$x_{r_1} = P_{f1}\bar{x}_{11}q_1(t) + P_{f2}\bar{x}_{12}q_2(t) \tag{6.40}$$

$$x_{r_2} = P_{f1}\bar{x}_{21}q_1(t) + P_{f2}\bar{x}_{22}q_2(t) \tag{6.41}$$

or by SRSS

$$x_{r_1,m} = \sqrt{(P_{f1}\bar{x}_{11}q_{1,m})^2 + (P_{f2}\bar{x}_{12}q_{2,m})^2} \tag{6.42}$$

$$x_{r_2,m} = \sqrt{(P_{f1}\bar{x}_{21}q_{1,m})^2 + (P_{f2}\bar{x}_{22}q_{2,m})^2} \tag{6.43}$$

The modal maxima $q_{1,m}$ and $q_{2,m}$ can be obtained from respective response spectra plots. It could be sinusoidal base motion or earthquake ground motion of any other form of pulse or any arbitrary type ground motion.

6.3 SOLVED PROBLEMS

Example 1. A steel frame shown in Fig. 6.4a below is subjected to a horizontal force applied at a girder level. The force decreases linearly from 25 kN at time $t = 0$ to zero at $t = 0.7$ sec (Fig. 6.4b). Determine (a) horizontal deflection at $t = 0.5$ sec, and (b) maximum horizontal deflection. Assume columns are massless and girder is rigid. Neglect damping. The moment of inertia of column cross sections as $I = 28000$ cm^4.

Solution:

Equivalent stiffness k_{eq}

$$k_{eq} = \frac{12EI}{L_1^3} + \frac{3EI}{L_2^3}$$

$$= \frac{12 \times 2 \times 10^5 \times 280 \times 10^6}{(4500)^3} + \frac{3 \times 2 \times 10^5 \times 280 \times 10^6}{(4650)^3} = 9045.38 \, N/mm$$

The natural frequency is given by,

$$\omega = \sqrt{\frac{k_e}{m}} = \sqrt{\frac{9045.38 \times 10^3}{90000}} = 10 \, rad/sec$$

Modal Analysis

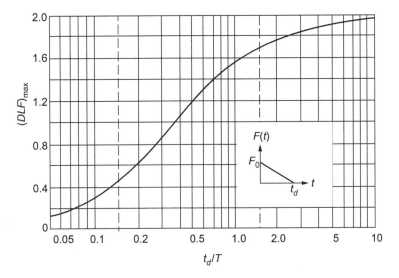

FIGURE 6.3: Max. dynamic load factor $(DLF)_{Max}$ for the undamped oscillator acted upon by a triangular force

Time period, $T = \frac{2\pi}{\omega} = \frac{2 \times 3.141}{\omega} = 0.63$ sec. $T < t_d$, therefore horizontal deflection at $t = 0.5$ sec

$$x(t) = \frac{\overline{F}}{k}(1 - \cos\omega t) + \frac{\overline{F}}{kt_d}\left(\frac{\sin\omega t}{\omega} - t\right)$$

$$\begin{aligned} x(0.5) &= \frac{25000}{9045.38}(1 - \cos(10 \times 0.5)) + \frac{25000}{9045.38 \times 0.7}\left(\frac{\sin(10 \times 0.5)}{10} - 0.5\right) \\ &= -1.92 \text{ mm} \end{aligned}$$

(b) Maximum horizontal deflection using Fig. 6.3.

$T = 0.63$ sec
For,

$$\frac{t_d}{T} = \frac{0.7}{0.63} = 1.11 \Rightarrow (DLF)_{max} \cong 1.6$$

We know that,

dynamic load factor is defined as the displacement at any time t divided by the static displacement $x_{st} = \frac{F}{k}$.

$$(DLF)_{max} = \frac{x_{max}}{x_{st}} \Rightarrow x_{max} = \frac{1.6 \times 25000}{9045.38} = 4.42 \text{ mm}$$

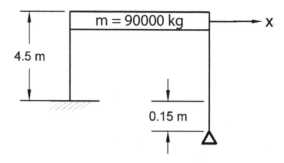

FIGURE 6.4a: The building frame

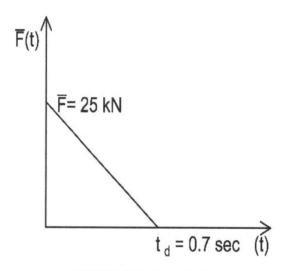

FIGURE 6.4b: Load details

Example 2. A two-storey frame as shown in Fig. 6.5 is to be designed for blast resistance. The blast at a distance from the frame is supposed to exert overpressure which is in the form of triangular pulse as shown in Fig. 6.6. The free vibration results are to be shown. Take $EI = 9 \times 10^{13}$ N-mm².

Solution:
The stiffness of each storey

$$k_1 = \frac{12EI}{L^3} \times 2 = \frac{12 \times 9 \times 10^{13}}{(3500)^3} \times 2 = 50380 \; N/mm$$

To find frequency,

$$[k - \omega^2 m] = 0$$

Modal Analysis

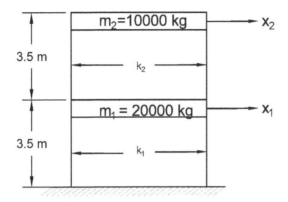

FIGURE 6.5: Two-storey frame

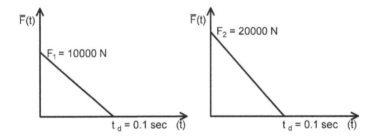

FIGURE 6.6: Blast load details

$$\begin{bmatrix} k_1+k_2 & -k_2 \\ -k_2 & k_2 \end{bmatrix} - \omega^2 \begin{bmatrix} m_1 & 0 \\ 0 & m_2 \end{bmatrix} = 0$$

Substituting values,

$$\begin{bmatrix} 100760 & -50380 \\ -50380 & 50380 \end{bmatrix} - \omega^2 \begin{bmatrix} 20000 & 0 \\ 0 & 10000 \end{bmatrix} = 0$$

On solving the above matrices, we get

$$\omega_1^2 = 1.44 \, rad^2/sec^2, \, \omega_2^2 = 8.9 \, rad^2/sec^2$$

$$\omega_1 = 1.2 \, rad/sec, \, \omega_2 = 2.98 \, rad/sec$$

To find mode shapes, solve the equations below:

$$(100760 - 20000\omega^2)\bar{x}_{11} - 50380\bar{x}_{21} = 0 \quad (a)$$

$$-50380\bar{x}_{12} + (50380 - 10000\omega^2)\bar{x}_{21} = 0 \quad (b)$$

In Equation (a), substitute $\omega^2 = \omega_1^2 = 1.44$ and $\bar{x}_{21} = 1$. In Equation (b), substitute $\omega^2 = \omega_2^2 = 8.9$ and $\bar{x}_{22} = 1$.

The following mode shapes are obtained:

$$\Phi_1 = \begin{bmatrix} \bar{x}_{11} \\ \bar{x}_{21} \end{bmatrix} = \begin{bmatrix} 0.7 \\ 1 \end{bmatrix}$$

$$\Phi_2 = \begin{bmatrix} \bar{x}_{12} \\ \bar{x}_{22} \end{bmatrix} = \begin{bmatrix} -0.766 \\ 1 \end{bmatrix}$$

Blast load details are

$$F_1(t) = 10000(1 - \frac{t}{t_d})N, F_2(t) = 20000(1 - \frac{t}{t_d})N; for\ t \leq t_d$$

Here $t_d = 0.1$ sec
$F_1(t)$ and $F_2(t) = 0$; for $t > t_d$
Time period of the frame is

$$T_1 = 5.2 \text{ sec.}$$
$$T_2 = 2.1 \text{ sec.}$$

Here both T_1 and T_2 are $> t_d$
To get the responses, substitute the values in the uncoupled equations below:

$$\bar{m}_1 \ddot{z}_1 + \omega_1^2 \bar{m}_1 z_1 = \bar{F}_1(t)$$

$$\bar{m}_2 \ddot{z}_2 + \omega_2^2 \bar{m}_2 z_2 = \bar{F}_2(t)$$

Here,

$$\bar{F}_1(t) = \bar{x}_{11} F_1(t) + \bar{x}_{21} F_2(t)$$

$$\bar{F}_2(t) = \bar{x}_{12} F_1(t) + \bar{x}_{22} F_2(t)$$

$$\bar{F}_1(t) = (0.7 \times 10000) + (1 \times 20000) = 27000\ kN = 27\ kN$$

$$\bar{F}_2(t) = (-0.76 \times 10000) + (1 \times 20000) = 12400\ N = 12.4\ kN$$

and

$$\bar{m}_1 = (m_1 x_{11}^2 + m_2 x_{21}^2)$$

$$\bar{m}_2 = (m_1 x_{12}^2 + m_2 x_{22}^2)$$

Modal Analysis

$$\overline{m}_1 = (20000 \times 0.7 \times 0.7) + (10000 \times 1 \times 1) = 19.8 \ kN$$

$$\overline{m}_2 = (20000 \times -0.76 \times 0.76) + (10000 \times 1 \times 1) = 21.55 \ kN$$

Since, $T > t_d$ the maximum values for z_1 and z_2 are obtained from a response spectra chart.

$$\frac{t_d}{T_1} = \frac{0.1}{5.2} = 0.02, \frac{t_d}{T_2} = \frac{0.1}{2.1} = 0.047$$

From the response chart of the triangular load as shown in Fig. 6.3,

$$(DLF)_{max,1} \cong 0.1, \ and \ (DLF)_{max,2} \cong 0.2$$

$$(DLF)_{max,1} = \frac{z_{1,max}}{z_{st,1}}, \ and \ (DLF)_{max,2} = \frac{z_{2,max}}{z_{st,2}}$$

$$\overline{k}_1 = \omega_1^2 \overline{m}_1 = 1.44 \times 19.8 = 28.51 \ kN/m$$

$$\overline{k}_2 = \omega_2^2 \overline{m}_2 = 8.9 \times 21.55 = 191.8 \ kN/m$$

Static displacement is

$$z_{st,1} = \frac{\overline{F}_1}{\overline{k}_1} = \frac{27}{28.51} = 0.94 \ m$$

$$z_{st,2} = \frac{\overline{F}_2}{\overline{k}_2} = \frac{12.4}{191.8} = 0.064 \ m$$

$$\Rightarrow z_{1,max} = 0.1 \times 0.94 = 0.094 \ m$$

$$\Rightarrow z_{2,max} = 0.2 \times 0.064 = 0.0128 \ m$$

Where, $z_{1,m}$ and $z_{2,m}$ are modal maxima. Now, we need to find out the actual values of max. storey displacement which are obtained after multiplying the $z_{1,m}$ and $z_{2,m}$ by respective components of eigenvector or modal vectors.

$$x_{1,m} = |\overline{x}_{11} z_{1,max}| + |\overline{x}_{12} z_{2,max}| = |0.7 \times 0.094| + |(-0.766 \times 0.0128| = 0.075 \ m$$

$$x_{2,m} = |\overline{x}_{21} z_{1,max}| + |\overline{x}_{22} z_{2,max}| = |1 \times 0.094| + |(1 \times 0.0128| = 0.1068 \ m$$

Above are very conservative estimates.

More reasonable and probable estimates are

$$x_{1,m} = \sqrt{(\bar{x}_{11}z_{1,m})^2 + \bar{x}_{12}z_{2,m})^2} = \sqrt{(0.7 \times 0.094)^2 + ((-0.766) \times 0.0128)^2} = 0.079 \ m$$

$$x_{1,m} = \sqrt{(\bar{x}_{11}z_{1,m})^2 + \bar{x}_{12}z_{2,m})^2} = \sqrt{(1 \times 0.094)^2 + (1 \times 0.0128)^2} = 0.095 \ m$$

Example 3. Table 3.1 shows the details of the structure. Find the response of the structure subjected to blast load.

Evaluating eigenvalues and eigenvectors
Mass matrix [M] and stiffness matrix [K] of the lumped mass model are

$$[M] \begin{bmatrix} m_1 & 0 & 0 \\ 0 & m_2 & 0 \\ 0 & 0 & m_3 \end{bmatrix} = \begin{bmatrix} 302 & 0 & 0 \\ 0 & 302 & 0 \\ 0 & 0 & 118.4 \end{bmatrix} (kN)$$

Table 3.1 Details of the structure

1		Type of building	G+2 1-Bay
2		Type of soil	Medium (II)
3		Floor height	3m
4		LL	3kN/m²
5		Top floor load	3kN/m²
6		Col. Size	400mm x 400mm
7		Beam Size	200mm x 450mm
8		Slab thickness	125mm
9		Importance factor	1
10		Response reduction factor	5
11		Earthquake zone (Z)	III (0.24)
12		Concrete Grade	M25
13		Thickness of infill	200mm
14		Infill density	20kN/m³
15		RCC density	25kN/m³

Modal Analysis

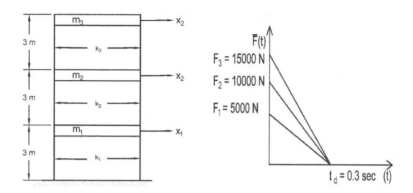

FIGURE 6.7: The three storeyed frame and the blast loading

In the table showing details of the structure generally all the details like earthquake zone, importance factor and response reduction factor are given for completeness. But here they are not relevant as the frame is designed for blast loads. Of course we do design a structure for ductility even in the case of blast loads as the structure invariably will go in to the inelastic range during the effect of blasts. But those details are not covered in the present book as the emphasis is only on the structural dynamics in earthquake and blast design.

Stiffness k $= \frac{12EI}{l^3} = 23{,}703.7$ kN-m $k_1 = k_2 = k_3 = 2 \times k = 47{,}407.4$ kN-m

$$[K] \begin{bmatrix} k_1+k_2 & -k_2 & 0 \\ -k_2 & k_2+k_3 & -k_3 \\ 0 & -k_3 & k_3 \end{bmatrix} = \begin{bmatrix} 94814.8 & -47407.41 & 0 \\ -47407.41 & 94814.8 & -47407.41 \\ 0 & -47407.41 & 47407.41 \end{bmatrix} kN-m$$

To obtain eigenvalues and eigenvectors we solve the matrices:

[k - λ M] = 0;

Where $\lambda = \omega^2$ = natural frequency.

By solving, natural frequencies of various modes are

$$[\omega^2] = \begin{bmatrix} 21.97 & 0 & 0 \\ 0 & 168.4 & 0 \\ 0 & 0 & 321.8 \end{bmatrix}$$

The following equations are used to evaluate frequency in each mode:

$$[K - \omega_1^2 M]\bar{x}_1 = 0;$$

$$[K - \omega_2^2 M]\bar{x}_2 = 0;$$

$$[K - \omega_3^2 M]\bar{x}_3 = 0.$$

On solving the equations, mode shapes (eigenvectors) are obtained.
Eigenvectors $\bar{x} = [\bar{x}_1 \ \bar{x}_2 \ \bar{x}_3] =$

$$\begin{bmatrix} 1 & 1 & 1 \\ 1.72 & -0.15 & -2.1 \\ 1.93 & -0.92 & 3.45 \end{bmatrix}$$

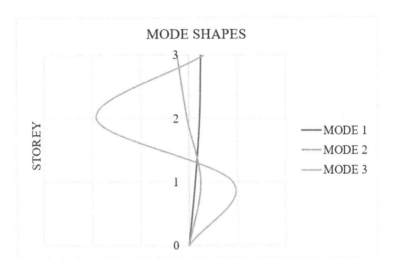

Natural time period $T = \frac{2\pi}{\omega}$

$$T = \begin{bmatrix} 1.341 & 0 & 0 \\ 0 & 0.48 & 0 \\ 0 & 0 & 0.35 \end{bmatrix} sec$$

Blast load details are

$$F_1(t) = 5000(1 - \frac{t}{t_d})N, F_2(t) = 10000(1 - \frac{t}{t_d})N; for T \leq t_d$$

$$F_3(t) = 15000(1 - \frac{t}{t_d})N; for T \leq t_d$$

Here $t_d = 0.3$ sec
F1(t) and F2(t) = 0; for T > t_d
Here T > t_d
To get the responses, substitute the values in the uncoupled equations below:

$$\bar{m}_1 \ddot{z}_1 + \omega_1^2 \bar{m}_1 z_1 = \bar{F}_1(t)$$

$$\bar{m}_2 \ddot{z}_2 + \omega_2^2 \bar{m}_2 z_2 = \bar{F}_2(t)$$

$$\bar{m}_3 \ddot{z}_3 + \omega_3^2 \bar{m}_2 z_3 = \bar{F}_3(t)$$

Modal Analysis

Here,

$$\bar{F}_1(t) = \bar{x}_{11}F_1(t) + \bar{x}_{21}F_2(t) + \bar{x}_{31}F_3(t)$$

$$\bar{F}_2(t) = \bar{x}_{12}F_1(t) + \bar{x}_{22}F_2(t) + \bar{x}_{32}F_3(t)$$

$$\bar{F}_3(t) = \bar{x}_{13}F_1(t) + \bar{x}_{23}F_2(t) + \bar{x}_{33}F_3(t)$$

$$\bar{F}_1(t) = (1 \times 5000) + (1.72 \times 10000) + (1.93 \times 15000) = 51150\ N$$

$$\bar{F}_2(t) = (1 \times 5000) + (-0.15 \times 10000) + (-0.92 \times 15000) = -10300\ N$$

And,

$$\bar{F}_3(t) = (1 \times 5000) + (-2.1 \times 10000) + (3.45 \times 15000) = 35750\ N$$

$$\bar{m}_1 = (m_1\bar{x}_{11}^2 + m_2\bar{x}_{21}^2 + m_3\bar{x}_{31}^2)$$

$$\bar{m}_2 = (m_1\bar{x}_{12}^2 + m_2\bar{x}_{22}^2 + m_3\bar{x}_{32}^2)$$

$$\bar{m}_3 = (m_1\bar{x}_{13}^2 + m_2\bar{x}_{23}^2 + m_3\bar{x}_{33}^2)$$

$$\bar{m}_1 = (302 \times 1 \times 1) + (302 \times 1.72 \times 1.72) + (118.4 \times 1.91 \times 93) = 1636.47\ kg$$

$$\bar{m}_2 = (302 \times 1 \times 1) + (302 \times 0.15 \times 0.15) + (118.4 \times 0.92 \times 0.92) = 409\ kg$$

$$\bar{m}_3 = (302 \times 1 \times 1) + (302 \times 2.1 \times 2.1) + (118.4 \times 3.45 \times 3.45) = 5228.38\ kg$$

Since, T > t_d the maximum values for z_1 and z_2 are obtained from a response spectra chart

$$\frac{t_d}{T_1} = \frac{0.3}{1.341} = 0.223, \frac{t_d}{T_2} = \frac{0.3}{0.48} = 0.625, \frac{t_d}{T_2} = \frac{0.3}{0.35} = 0.85 \quad (6.44)$$

From the response chart of the triangular load as shown in Fig. 6.2.

$$(DLF)_{max1} \cong 0.7, (DLF)_{max2} \cong 1.3, (DLF)_{max3} \cong 1.5 \quad (6.45)$$

w.k.t

$$(DLF)_{max1} = \frac{z_{1max}}{z_{st1}}, (DLF)_{max2} = \frac{z_{2max}}{z_{st2}}, (DLF)_{max3} = \frac{z_{3max}}{z_{st3}} \quad (6.46)$$

$$\bar{k}_1 = \omega_1^2 \bar{m}_1 = 22.63 \times 1636.47 = 37033.2 \ N/m$$

$$\bar{k}_2 = \omega_2^2 \bar{m}_2 = 168.4 \times 409 = 68875.5 \ N/m$$

$$\bar{k}_3 = \omega_3^2 \bar{m}_3 = 321.8 \times 5228.38 = 1682492.7 \ N/m$$

Static displacement is

$$z_{st1} = \frac{\bar{F}_1}{\bar{k}_1} = \frac{51150}{37033.2} = 1.38 \ m$$

$$z_{st2} = \frac{\bar{F}_2}{\bar{k}_2} = \frac{-10300}{68875.6} = 0.15 \ m$$

$$z_{st3} = \frac{\bar{F}_3}{\bar{k}_3} = \frac{35750}{1682492.7} = 0.02 \ m$$

$$\Rightarrow z_{1max} = 1.38 \times 0.7 = 0.966 \ m$$

$$\Rightarrow z_{2max} = 0.15 \times 1.3 = 0.195 \ m$$

$$\Rightarrow z_{3max} = 1.5 \times 0.02 = 0.03 \ m$$

Where, $z_{1,m}$ and $z_{1,m}$ are modal maxima. Now, we need to find out the actual values of max. storey displacement which are obtained after multiplying the $z_{1,m}$ and $z_{1,m}$ by the respective components of vectors or modal vectors.

$$x_{1,m} = |\bar{x}_{11} z_{1,max}| + |\bar{x}_{12} z_{2,max}| + |\bar{x}_{13} z_{3,max}| \tag{6.47}$$

$$= |1 \times 0.966| + |1 \times 0.195| + |1 \times 0.03| = 1.191 \ m$$

$$x_{2,m} = |\bar{x}_{21} z_{1,max}| + |\bar{x}_{22} z_{2,max}| + |\bar{x}_{23} z_{3,max}| \tag{6.48}$$

$$= |1.72 x 0.966| + |0.15 \times 0.1950| + |2.1 x 0.03| = 1.75 \ m$$

$$x_{3,m} = |\bar{x}_{31} z_{1,max}| + |\bar{x}_{32} z_{2,max}| + |\bar{x}_{33} z_{3,max}| \tag{6.49}$$

$$= |1.93 \times 0.966| + |0.92 \times 0.195| + |3.45 \times 0.03| = 2.15 \ m$$

Modal Analysis

Above are very conservative estimations.

More reasonable and probable estimations are

$$x_{1,m} = \sqrt{(\bar{x}_{11}z_{1,m})^2 + (\bar{x}_{12}z_{2,m})^2 + (\bar{x}_{13}z_{3,m})^2} \tag{6.50}$$

$$x_{1,m} = \sqrt{(1 \times 0.966)^2 + (1 \times 0.195)^2 + (1 \times 0.03)^2} = 0.98 \, m \tag{6.51}$$

$$x_{2,m} = \sqrt{(\bar{x}_{21}z_{1,m})^2 + (\bar{x}_{22}z_{2,m})^2 + (\bar{x}_{23}z_{3,m})^2} \tag{6.52}$$

$$= \sqrt{(1.72 \times 0.966)^2 + (0.15 \times 0.195)^2 + (2.1 \times 0.03)^2} = 1.66 \, m$$

$$x_{3,m} = \sqrt{(\bar{x}_{31}z_{1,m})^2 + (\bar{x}_{32}z_{2,m})^2 + (\bar{x}_{33}z_{3,m})^2} \tag{6.53}$$

$$= \sqrt{(1.93 \times 0.966)^2 + (0.92 \times 0.195)^2 + (3.45 \times 0.03)^2} = 1.87 \, m$$

Example 4. Determine the responses of the frame shown in Example 2, when it is subjected to a suddenly applied constant load acceleration $\ddot{x}_S = 0.15$ g at its base. The natural frequency and corresponding modes from the calculations in Example 2 are

$$\omega_1 = 1.2 \, rad/sec, \omega_2 = 2.98 \, rad/sec$$

$$\bar{x}_{11} = 0.7, \bar{x}_{12} = -0.766$$

$$\bar{x}_{21}, \bar{x}_{22} = 1$$

Solution:

The acceleration acting at the base of the structure is

$$\ddot{x}_s = 0.5 \times 9.81 = 1.47 \, m/sec^4$$

The participation factors are calculated.

$$P_{f1} = -\frac{m_1 \bar{x}_{11} + m_2 \bar{x}_{21}}{m_1 \bar{x}_{11}^2 + m_2 \bar{x}_{21}^2} = \frac{(20000 \times 0.7) + (10000 \times 1)}{(20000 \times 0.7^2) + (10000 \times 1^2)} = -1.21 \tag{6.54}$$

$$P_{f2} = -\frac{m_1 \bar{x}_{12} + m_2 \bar{x}_{22}}{m_1 \bar{x}_{12}^2 + m_2 \bar{x}_{22}^2} = \frac{(20000 \times -0.76) + (10000 \times 1)}{(20000 \times -0.76^2) + (10000 \times 1^2)} = 0.24 \tag{6.55}$$

The modal equations are

$$\ddot{q}_1 + \omega_1^2 q_1 = \ddot{x}_s(t) \Rightarrow \ddot{q}_1 + 1.44 q_1 = 1.47$$

$$\ddot{q}_2 + \omega_2^2 q_2 = \ddot{x}_s(t) \Rightarrow \ddot{q}_2 + 1.44 q_2 = 1.47$$

The solution assuming initial conditions for velocity and displacement is

$$q_1(t) = \frac{1.47}{1.44}(1 - \cos 1.2 t)$$

$$q_2(t) = \frac{1.47}{8.9}(1 - \cos 2.98 t)$$

The response in terms of relative motion at floor levels with respect to displacement of the base is given by Equations 6.18 and 6.19

$$x_{r_1,m} = -1.21 \times 0.7 \times 1.02(1 - \cos 1.2 t) + 0.24 \times (-0.766) \times 0.165(1 - \cos 2.98 t)$$

$$x_{r_2,m} = -1.21 \times 1 \times 1.02(1 - \cos 1.2 t) + 0.24 \times 1 \times 0.165(1 - \cos 2.98 t)$$

Due to simple excitation function (constant acceleration), it was possible to obtain a closed solution as a function of time. For an actual earthquake, the response is obtained through numerical integration or response spectra if available.

The maximum modal response in the present example is obtained when cosine functions are equal to minus one.

Thus, the maximum modal response is then

$$q_{1,m} = 2.04$$

$$q_{2,m} = 0.33$$

Maximum response is calculated from approximate formulas (SRSS), i.e., Equation 6.20 and 6.21

$$x_{r_1,m} = \sqrt{(1.21 \times 0.7 \times 2.04)^2 + (0.24 \times (-0.766) \times 0.33)^2} = 1.73 \ m$$

$$x_{r_2,m} = \sqrt{(1.21 \times 1 \times 2.04)^2 + (0.24 \times 1 \times 0.33)^2} = 2.47 \ m$$

The possible maximum values for the responses are calculated from Equations 6.18 and 6.19 and by setting the cosine functions to their maximum value.

$$x_{r1,m} = 1.78 \ m$$

$$x_{r2,m} = 2.54 \ m$$

6.4 EXERCISE PROBLEMS FOR CHAPTERS 5 AND 6

(1) A two-storey frame as shown in Example 2 is subjected to a triangular load of 1 kN and 3.5 kN at the first and second storey, respectively. Determine the displacements at storey levels. Take $EI = 11 \times 10^{13}$ N-mm^2.

(2) Repeat Problem 1 for 15% critical damping.

(3) Take 3 bays for the frame shown in Example 2 and find out the force transmitted to the support, if the width of the bay is 4 m. Take damping ratio η as 0.07.

(4) Find out the response of a three-storey frame as shown in Example 3 if it is subjected to a ground acceleration of 0.15 g. Take 200 mm × 300 mm column and beams of 200 mm × 200 mm.

(5) In Example 2 compare the equivalent static analysis with those of response spectrum analysis and find out whether scale factor is required or not.

(6) Take ground storey height as 4.5 m in Example 3 and check whether it satisfies the soft storey condition as per IS 1893:2016.

(7) Take 4 bays for the frame shown in Example 3 and find out the displacement at storey levels if it is subjected to ground acceleration of 0.3 g. Take column dimension as 230 mm $\times \eta$ 375 mm, beam dimension as 230 mm × 230 mm, and slab thickness as 150 mm. Take live load as 2 kN/m^2. Assume no damping and width of the bay as 3.5 m.

(8) Repeat Problem 7 when it is subjected to a suddenly applied constant load acceleration 0.3 g at its base.

7 Earthquake Resistant Design

7.1 INTRODUCTION

Earthquake or seismic resistant design is an important aspect of the structural design. We discussed earlier single degree of freedom and multi-degree of freedom systems subjected to base motion. It will be very important now to understand the seismic response of structures. Earthquake ground motions are nothing but base motions. The base motion during earthquakes can be experienced as acceleration or velocity or displacement.

The recorders at seismological observatories will be automatically triggered to measure the ground motions as soon as they occur. They record all the above three parameters in all three directions, that is two horizontal and one vertical. The recorders can be situated either close to the epicentre or far away from it, depending on the geographical features. Therefore, there are both near field and far field ground motions recorded in the literature. The response of a structure is very different in the two cases and further, natural frequencies of the structure also determine whether it is more vulnerable to near field or far field ground motions.

There is enough information about geological features of ground motion already in various textbooks hitherto published, and therefore they will not be repeated here. Information on mantle, focus and epicenter and so on is plenty in the earlier textbooks and therefore the reader is expected to refer to the earlier books (in the list of references). Besides, these are what are called deep earthquakes.

It is not the intention here to discuss such details, as they are well documented. We will concentrate more on the structural dynamic analysis of structures subjected to earthquakes.

7.2 STRUCTURAL ANALYSIS

Structural analysis of structure subjected to earthquakes is done by three methods. The simplest is the equivalent static method, which does not involve any dynamics at all. The next improved method is the response spectrum method, which is a little more sophisticated and involves dynamics to some extent indirectly. The third and the most difficult procedure is dynamic analysis, which consists of time history analysis of the response in totality.

7.3 STRUCTURAL MODEL

In dynamic analysis, as stated above, any of the three methods, requires a simpler model of a real structure. If a real building structure has to be treated as it is in all its details, it is very complex. The nearest method to follow all the structural details is the well-known Finite Element Method (FEM). Even in FEM a continuous system such as beams, columns and slabs has to be discretized into smaller elements with masses and stiffnesses lumped at the nodes. Otherwise, if one has to represent a structure in its real form, one has to adopt very large number of elements. If beams, columns and even slabs have to be approached as continuous systems, which is extremely complex solving a sea of partial differential equations with complicated numerous boundary conditions is almost impossible.

Therefore, in order to overcome the above difficulties, we start with as many assumptions as possible, of course reasonable ones, and make the model as well as the analysis simpler.

7.4 SHEAR BUILDING

A moderately tall multi-storey building can be modelled with the following assumptions of beams and slabs:

(1) All masses due to self-weight, live load and all other transverse dead loads, self-weight of the columns, infills are assumed to be lumped at the floor levels.

(2) The floor slabs are considered to be rigid in their own plane.

(3) The stiffness of the building structure is provided only by the columns. The columns are assumed to be fixed in rotation at the floor levels.

(4) The degrees of freedom are defined at the floor and the roof levels only. Maximum number per floor or roof will be three, viz. two translations and one rotation (in case of asymmetry).

The above main assumptions make the analysis very simple, particularly the lateral stiffness of any storey can be very easily obtained and the degrees of freedom will be lesser as it depends on the number of floors.

For example, in a typical multi-storey building which has no asymmetry in plan (there is no rotation of the building in plan), and if the frame is assumed to undergo oscillations in its own plane, then there can be only lateral degrees of freedom in that plane. Therefore, as the degrees of freedom can be assigned at the masses, the total number of degrees of freedom will simply be the number of masses or in other words number of floors.

The main difference between static and dynamic analysis of structures is the inertial force (mass x acceleration) term. Static analysis considers force equilibrium as

F= kx, while in dynamic analysis force equilibrium as is F(t) = kx(t) + m$\ddot{x}(t)$, where x(t) is the time dependent displacement and $\ddot{x}(t)$ is the time dependent acceleration of the mass.

In equivalent static method, the base shear is distributed over the masses, of various floors over the height according to the 1st mode shape. The mass multiplied by the acceleration is considered as a static force acting on the mass. The frame is analyzed as if it is subjected static forces. Static analysis being a much simpler, equivalent static method is very popular among many practising engineers.

7.5 RESPONSE SPECTRUM

Response spectrum is a plot of maximum responses of a single degree of freedom system over various periods and for different values of damping when the system is subjected to the same input or, it can be defined as a plot of maximum responses for several single degree of freedom systems of different natural frequencies or different natural periods for a given input force (Anil K. Chopra 2001, Ray W. Clough and Joseph Penzien, 1982). Damping is held constant for each plot. The input could be sinusoidal impulse, pulse of different shapes and earthquake ground motion. The concept of response spectrum was introduced long ago. After M.A. Biot gave the concept in 1932, G.W. Housner found its application.

The maximum responses for a given input of earthquake will be useful to obtain the maximum response of a multi-degree of freedom system by the mode superposition technique. Another alternative is the more rigorous dynamic analysis following the time history of the earthquake as input and solving the coupled equation of motion, by numerical integration procedure, such as Newmark-Beta and Wilson-Theta methods.

A typical response spectrum for the El Centro 1940 ground motion is shown in Fig. 7.1. Generally El Centro ground motion is commonly used as an example in all books and publications because it is one of the very early recorded ground motions and also it caused a huge catastrophic damage in California, USA. The response spectrum plot shown is represented as a tripartite plot. The horizontal axis contains period T, of SDF system. Maximum pseudo-velocity $\dot{x}_{pv}$ is plotted on the vertical axis, while maximum pseudo-acceleration $\ddot{x}_{pa}$ and maximum deformation x_{rm} are along the 45° lines shown. It is on a logarithmic scale. Further, the displacement, velocity and acceleration values are normalized with respect to peak values of ground motion like peak ground acceleration, peak ground velocity and peak ground displacement.

If damping is varied, you will see plots with slightly different values, although, the overall shape is more or less similar. Observing the response spectrum plot (Fig. 7.1) it is possible think of an idealised form of the plot consisting of a few straight lines. The idealised plot will help us to draw certain conclusions. The idealised plot consisting of straight lines looks like a trapezium with broader side being the bottom one. It has two sloping lines on the left and the right sides with a flat portion at the top. Without

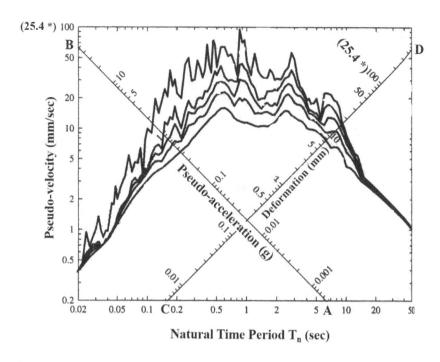

FIGURE 7.1: Tripartite plot of response spectra for El Centro ground motion; $\eta = 0, 2, 5, 10, 20\%$

any loss of generality it can be assumed to represent the response spectrum plot. The sloping line on the left will intersect the 45 degree line at acceleration levels larger than the peak ground acceleration (Fig. 7.4). That will be from $T = 0.035$ sec to $T = 0.5$ sec. They are short period structures. The system peak pseudo acceleration exceeds the peak ground acceleration. However it will be assumed that it is equal to the peak ground acceleration. In the middle from $T = 0.5$ sec to 3.0 sec the pseudo velocity will be almost constant equal to the peak ground velocity while from about $T = 3.0$ sec to 15 sec, the sloping line on the right side the maximum relative displacement of the system will be equal to the peak ground displacement. However, between $T_b = 0.125$ sec and $T_c = 0.5$ sec, the value of $\ddot{x}_{sm}/\ddot{x}_{gm}$ can be considered as almost constant equal to unity, or in other words the peak pseudo-acceleration is equal to peak ground acceleration. Similarly, between $T_d = 3.0$ sec and $T_f = 15.0$ sec is essentially a long period system or a flexible system, the maximum relative displacement x_{rm} exceeds the peak ground displacement x_{gm}. The amplification depends again on both T and damping η. However, between $T_d = 3.0$ sec and $T_e = 10.0$ sec, the maximum relative displacement can be considered to be almost constant equal to maximum ground displacement. On the same lines between $T = 0.5$ sec and $T = 3.0$ sec, the maximum pseudo-velocity $\dot{x}_{sm}$ appears to be almost constant equal to peak ground velocity $\dot{x}_{gm}$. Based on the above observations, the response spectrum can be broadly

Earthquake Resistant Design

divided into three regions, with T<T_c, between T_c and T_d and > T_d. In short it will be nice to understand the following, for very short period or very stiff systems, the mass moves along with the ground. Let us imagine a short cantilever pole. It moves the same distance as the ground below, which means the acceleration of the pole is the same as that of the ground (the relative acceleration is zero) and the relative displacement $x_r = 0$, or the absolute displacement x = ground displacement x_g. Let us see the example of a very flexible system or a long period one. It is so flexible that the mass remains practically stationary, while the ground below moves. Therefore, the relative displacement, x_r = ground displacement x_g. Because the mass is particularly at rest, the acceleration and thus the inertia force is zero.

However, it may be worth repeating that finally we will not make use of response spectrum for any particular earthquake ground motion, because it is never possible to know which earthquake, of what peak ground acceleration, of what frequency content and so on occurs at a particular site. Therefore, the average spectra as suggested by Housner is generally useful in design and IS 1893-2016 also contains an average spectrum. It suits any earthquake in an average sense.

However, the Response Spectrum Method, although not as accurate as the above mentioned rigorous dynamic analysis, is more handy and useful in the design of multi-storey structures for earthquakes, because we are interested only in the maximum values.

Having understood that R-S is a plot of maximum responses vs. period or also maximum responses vs. natural frequency, we now see how we plot the same. Let us also realize that in literature and all codes of practice, a period is used rather than natural frequency as the parameter on the horizontal axis in the R-S plot.

One of the straightforward ways of obtaining the maximum response is to write the equation of motion of a single degree of freedom system where mass, damping and stiffness are well defined. The system is subjected to a known force, may be sinusoidal or a pulse of any known shape or earthquake ground (support) motion. Here let us concentrate on single degree of freedom system subjected to a support motion, say acceleration whose time history is known.

For example, let us discuss ground motion of a known earthquake. Very popularly, the earthquake which occurred in El Centro, California in 1940 is considered as the standard data by all the researchers in the past, because that is the earthquake which caused a huge disaster (Fig. 7.2).

The peak amplitude was 0.33 g which means a building will be subjected to a lateral force which is about 33% of its weight. A human being generally can carry several times his own self-weight, for example weight lifters carry as high as 200-300 kg, while the same human being finds it difficult to stand a lateral force of (if someone pulls or pushes him about 20-30% of his own weight, unless his legs are specially trained like those of a soccer world champion). A human being with less mass of the upper body with much stronger pair of legs could be a good soccer player. It is in the

E1 Centro 1940 - S00E

FIGURE 7.2: Ground acceleration record

same way, columns of buildings have to be much stronger with possible less vertical load in the form of dead load and live load.

The equation of motion with right hand side being the time history input is integrated numerically (because the input is not in the deterministic time frame) to obtain the time history of displacements, velocity and accelerations of the system. The maximum value, viz. displacement, velocity and acceleration, is predicted from the time history is plotted against the natural period of the single degree of freedom system. (for the same input, the period of the system is varied by varying either the mass or stiffness or both, and the maximum values are picked up from the output time history response).

Thus the maximum values are next plotted against the new period. The procedure is repeated to obtain the maximum values of displacement, velocity and acceleration of the corresponding period of the SDF system. The maxima as obtained above are real, precisely determined maximum values. However, there is yet another easier but approximate way of obtaining the max values of velocity and acceleration from max relative displacement which are obtained from solving the equation of motion the max velocity and acceleration are obtained by multiplying the relative displacement x_r by ω and ω^2, respectively, where ω is natural frequency. The max values thus obtained are called pseudo-velocity and pseudo-acceleration because they are not the true values as determined earlier.

The details are as follows:

The equation of motion of a SDF subjected to the support motion $x_s(t)$ is (Fig. 7.3)

$m\ddot{x} + c\ddot{x}_r + kx_r = 0$
Where $x_r = x - x_s$ is the relative displacement,
m, k and c are mass, stiffness and damping constants, respectively.

Let us assume that c=0,
then, $m\ddot{x} + kx_r = 0$
$\ddot{x} = -\omega^2 x_r$
or even if c $\neq$ 0,
$\ddot{x}_m = -\omega^2 x_{rm}$ because at maximum displacement $\dot{x} = 0$ and therefore $c\dot{x} = 0$. Note that $\ddot{x}_m$ is the maximum absolute acceleration and x_{rm} is the maximum relative displacement.

Earthquake Resistant Design

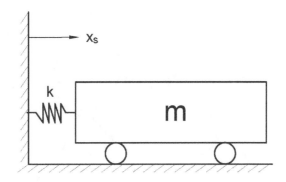

FIGURE 7.3: SDF subjected to support motion

The max acceleration $\ddot{x}_{pa}$ is termed as spectral or pseudo-acceleration.

As mentioned above,

$\ddot{x}_m = -\omega^2 x_{rm}$, even when $c \neq 0$

$= \omega^2 x_{rm}$ neglecting the negative sign, as it is not relevant.

As seen above, even when $c \neq 0$, the max displacement occurs when $\dot{x} = 0$, thus the middle term vanishes and thus the same argument holds good. It shows, $\ddot{x}_m$ is related to x_{rm} through ω.
Following the above argument $\dot{x}_{pv} = \omega x_{rm}$ is called spectral or pseudo-velocity.

By a strange coincidence, the above formulation looks similar if the input were to be harmonic.
$x = \bar{x} \sin \omega t$
$\dot{x} = \omega \bar{x} \cos \omega t$
$\ddot{x} = -\omega^2 \bar{x} \sin \omega t$
and the respective maximum values are
$x_{rm} = \bar{x}$
$\dot{x}_{pv} = \omega \bar{x}$
$\ddot{x}_{pa} = -\omega^2 \bar{x}$

It may be noted that the maximum values obtained from the way above by multiplying maximum relative displacement by ω and ω^2 are not the true values obtained from the time history. Therefore, they are called pseudo-velocity and pseudo-acceleration, $\dot{x}_{pv}$ and $\ddot{x}_{pa}$, respectively. They are also termed as spectral values.

Therefore in all our future response spectrum plots we employ pseudo-acceleration and pseudo-velocity as the parameters because they can be obtained easily. They are

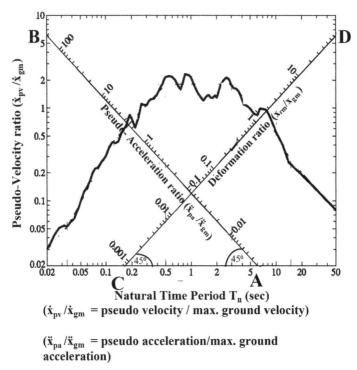

$(\dot{x}_{pv}/\dot{x}_{gm}$ = pseudo velocity / max. ground velocity)

$(\ddot{x}_{pa}/\ddot{x}_{gm}$ = pseudo acceleration/max. ground acceleration)

$(x_{rm}/x_{gm}$ = max. relative displacement/max. ground displacement)

FIGURE 7.4: Response spectra for El Centro ground motion shown by a solid line together with an idealized version shown by a dashed line $\eta = 5\%$

sufficiently close to the true values determined from the numerical integration of the equation of motion.

It is observed from the damages caused in buildings after several earthquakes, in the past, that the base shear to which the buildings are subjected are much larger than the code specified values for which the buildings were designed. That revealed that those buildings had dissipated a sufficient portion of the energy of the earthquake through yielding mechanism of the structural members. That further led to the importance of yielding in structural members and design for inelastic deformation. The concept of ductility emerged therein.

The inelastic response spectrum is a spectrum which gives maximum ductility (ratio of max deflection to yield deflection) caused due to a certain earthquake over various values of fundamental periods of a SDF. Most of the previous investigators employed time history of El Centro 1940 for the spectrum (Fig. 7.2). The max ductility

Earthquake Resistant Design

is plotted against the period of the structure. Veletsos has worked extensively on such spectra. They are called inelastic response spectra.

Following the preceding instance cited above, it can be said that the base shear in buildings designed according to the code provisions is less and sometimes farless than the base shears generated, during the earthquake, ofcourse depending on the magnitude of the earthquake. It may be questioned now, why not design the buildings for much larger shear based on the previous experience. The answer here is very nicely stated in the limit state philosophy in earthquake design. The philosophy is that it is prudent to design buildings for moderate and less than moderate magnitudes, such that they remain within elastic limits. When we say moderate, which is a subjective term, we mean that these earthquakes have a smaller return period. Similarly, based on the same logic, when structures are designed for earthquakes of larger magnitudes which automatically mean earthquakes of longer return period, it is wise to design such that some parts of the structures yield and thus expend earthquake energy. However, when the members yield they should not at any cost fail and collapse, resulting in injury to people and loss of lives. Here comes the concept of ductility. If the members have adequate capacity for ductile deformations, then they do not collapse. Therefore, in the inelastic design, we design for ductility. Hence the need for inelastic response spectrum.

We will take the example of the typical accelerogram. Generally one chooses the example of El Centro 1940 earthquake.

If a structure has to be designed for any specific earthquake, then it is much easier if one generates response spectrum for that particular earthquake and adopts it for the design. But the question is how to predict what kind of earthquake is going to occur in the future. Therefore, Housner thought of generating an average response spectrum which can be used as design spectra. Let us see how we generate average design spectra.

To draw a design spectrum which is an average spectrum, which should be good and reliable for design, one needs to draw spectra for several real earthquakes which occurred in the past, and perform a statistical analysis to obtain mean and standard deviation, σ to draw further spectra for 1σ, 2σ and 3σ. The spectra this drawn will be further normalised with respect to some other parameter like peak acceleration.

Method to draw a design spectrum

As earlier mentioned, the design spectrum is obviously an average spectrum, the average being that over several earthquake responses which have occurred previously. Naturally to get a good average spectrum, there should be a large number of earthquake ground motions recorded previously whose effects have been observed and measured. The only two countries in the world where the very large number of earthquakes have occurred are the U.S. and Japan. Even then, the recorded earthquakes are less than those occurred because precise recording started only in 1934 in the U.S. Since then there are a good number of recorded earthquake ground motions. In Fig. 7.1 there are several spectra for different damping values, all for the same real earthquake ElCentro 1940. One can draw such for different earthquakes as many as possible using available records. Then taking mean and standard deviation values by a statistical analysis one can also draw mean and standard deviation spectra. Spectra for different percentiles are also possible by considering mean plus several standard deviation values. Depending on the probability values, one can choose the design spectrum.

How do we make use of design spectrum?

Having obtained the design spectrum, we can make use of it for the response spectrum analysis. In the R-S analysis, we make use of the well-known method called summation of modal responses. We have already come to know that any structure or a system in general when excited by a force vibrates or responds with all the modes combined. Therefore, an easier method of getting the responses of a structure to a given ground motion is to know its modal responses after multiplying them by a factor called participating factor which tells the contribution for each of the modes, the largest being from the first mode and reducing for higher modes. Then the question is how many modes we could consider in the total summation. It is certainly an important question. Earlier, the concept was for shear buildings, where the assumption is that the masses are lumped at the floor levels. The columns are considered to be fixed against rotation. When such shear buildings are subjected to earthquake ground motion, the first three modes are sufficient enough in the summation process. In such shear buildings subjected to horizontal earthquake ground motions, only along horizontal degrees of freedom one per floor in each direction would be sufficient. In dealing with multi-degree of freedom system the concept of modal mass comes into focus. Each mode being a SDF will have an equivalent mass and equivalent stiffness. The equivalent mass represents all the masses of the frame, is the mass of the equivalent single degree of freedom and is called modal mass of that mode. Similarly, we have equivalent stiffness representing all the different spring stiffnesses or column stiffnesses (or also called floor stiffnesses). The equation of an equivalent single degree of freedom is

$m_n \ddot{x}_n + k_n x_n = F_n(t)$
$m_n = n^{th}$ modal mass
$k_n = n^{th}$ modal stiffness
$x_n = n^{th}$ modal displacement
$F_n = n^{th}$ modal force

As mentioned earlier, each of the modes is said to be orthogonal to any other modes; the property is called the well-known orthogonal property. Further, the orthogonality is either with respect to mass or stiffness as the case may be.

Earthquake Resistant Design

The eigenvalue problem can be obtained as follows:

Equation of motion of 'N' degrees of freedom in matrix form is given by

$$[m][\ddot{x}] + [c][\dot{x}] + [k][x] = F(t) \quad (7.1)$$

[m], [c] and [k] are mass, damping and stiffness matrices.

$$\text{Velocity vector} [\dot{x}] = \begin{bmatrix} \dot{x}_1 \\ \dot{x}_2 \\ \cdot \\ \cdot \\ \cdot \\ \dot{x}_N \end{bmatrix}, \text{ displacement vector } [x] = \begin{bmatrix} x_1 \\ x_2 \\ \cdot \\ \cdot \\ \cdot \\ x_N \end{bmatrix} \text{ Force vector } [F(t)] = \begin{bmatrix} F_1(t) \\ F_2(t) \\ \cdot \\ \cdot \\ \cdot \\ F_n(t) \end{bmatrix}$$

When the above system executes free vibration, the right-hand side is zero which is essentially simple harmonic motion; all the coordinates or components of the eigenvectors will have the same time function in any particular mode. It means that all the components will attain the same value at the same time. The motion is synchronous. Therefore

$$\{x\} = \{\bar{x}\}_n f(t)$$

Where $\bar{x}_i$ is the constant amplitude and $f(t)$ is the corresponding time function, same for all the elements of x(t). Substituting in Equation (7.1)

$$\ddot{f}(t)[m]\{\bar{x}\}_n + f(t)[k]\{\bar{x}\}_n = 0, \text{ in any } n^{th} \text{ mode.} \quad (7.2)$$

It is already known that in the free vibration the motion is a Simple Harmonic Motion and therefore

$$\ddot{f}(t) = \omega^2 f(t)$$

where ω is natural frequency of motion.

Having obtained the eigenvalues (natural frequency) and eigenvectors (mode shapes) from the free vibration analysis, we obtain what is called mode participation factor P_{fn} of mode n. (The subscript n indicates the mode number)

The factor P_{fn} indicates the amount of mode participation in the total response. It is a weighting factor. The participation factors P_{fn} decrease as the mode number increases, thus the largest value of P_{fn} is for the first mode P_{f1} and the least is P_{fm} for the last m^{th} mode. Of course the above is true provided that there are no two different modes which have the same frequency, or in other words, there are no degenerate modes.

To summarize the above procedure, we start with the equations of motion for a multi-degree of freedom system which are obviously coupled in both stiffness and mass. Very often in typical building frames, where we make the assumptions that masses are lumped at the floor levels and degrees of freedom are defined there only, rotations of the columns are neglected at the floor levels, we have the mass matrix which is diagonal, while the stiffness matrix is symmetric and square. Where the mass matrix is diagonal, the equations of motion are uncoupled in mass and coupled only in stiffness. In mode superposition method, the first step is to decouple the equations of motion both in mass and stiffness. In order to decouple, one makes use of orthogonality condition, which is

$$\{\bar{x}\}_n^T [m] \{\bar{x}\}_m = 0, where \quad n \neq m$$

$$\{\bar{x}\}_n^T [k] \{\bar{x}\}_m = 0 \tag{7.3}$$

The first one is orthogonality condition with respect to mass, and the second one is that with respect to stiffness.

Orthogonality condition as stated above can be proved to exist among eigenvectors either with respect to mass matrix or with respect to stiffness matrix. Therefore, to enable that, start with undamped equation of motion. Later, when we have to apply the orthogonality condition even for a damped system, we need to make certain assumptions like the damping is either mass or stiffness proportional or both. It will be like $[c] = c_1[k] + c_2[m]$ where c_1 and c_2 are some constants.

Now starting with the equations of motion of an undamped system, in matrix form,

$$[m]\{\ddot{x}\} + [k]\{x\} = 0$$

In free vibrations the motion is Simple Harmonic Motion and therefore, all the coordinates or masses will vibrate with the same frequency. The general shape of the vibration system remains the same.

$$\{x\} = \{\bar{x}\} Sin \omega t$$

On substitution, cancelling the time terms,

$$[k]\{\bar{x}\} = \omega^2 [m]\{\bar{x}\}$$

The above is called an Eigenvalue Problem: ω^2 and $\bar{x}$ are unknown.

Earthquake Resistant Design

The non-trivial solution, where $\bar{x} \neq 0$ exists only when the determinant of the matrix $|[k] - \omega^2[m]| = 0$. Expanding the determinant we obtain an algebraic equation called characteristic equation. It is called characteristic because it yields eigenvalues or natural frequencies and also eigenvectors or modal vectors, which are characteristics of the system. They are system properties depending on $[m], [k]$ and do not change with the time or external force. After obtaining the eigenvalues ω_n^2, substituting back in Equation 7.4, we obtain the values of $[\bar{x}]$ which we call eigenvectors which do not have absolute values but as relative values because there is no force acting on the system. The RHS of the equation of motion is zero. The value of the eigenvector can be normalized in a way according to convenience. The eigenvectors are represented by $\bar{x}_{in}$, where i stands for mass and n stands for mode number.

For example, in any n^{th} mode.

$$[k](\bar{x}_n) = \omega_n^2[m](\bar{x}_n) \tag{7.4}$$

ω_n is the frequency of the nth mode.

For example, the combined modal vector

$[\bar{x}] = [\{\bar{x}_1\}\{\bar{x}_2\} --- \{\bar{x}_n\}]$ where n = 1 to N modes.

Where $\{\bar{x}_n\}$ is any column vector in the matrix $[\bar{x}]$ of modal vectors of all modes.

Let us write the eigenvalue problem in any two distinct modes 'n' and 'm'

in n^{th} mode: $[k]\{\bar{x}\}_n = \omega_n^2[m]\{\bar{x}\}_n$......(a)

in m^{th} mode: $[k]\{\bar{x}\}_m = \omega_m^2[m]\{\bar{x}\}_m$......(b)

pre-multiplying both sides of the above Equation (a) by $\{\bar{x}\}_n^T$ and Equation (b) by $\{\bar{x}\}_n^T$

we obtain

$$\{\bar{x}\}_m^T[k](\{\bar{x}\}_n = \omega_n^2\{\bar{x}\}_m^T[m]\{\bar{x}\}_n$$

Similarly,

$$\{\bar{x}\}_n^T[k](\{\bar{x}\}_m = \omega_m^2\{\bar{x}\}_n^T[m]\{\bar{x}\}_m$$

subtracting one from the other after transposing only one of them,

$$(\omega_n^2 - \omega_m^2)\{\bar{x}\}_n^T[m]\{\bar{x}\}_m = 0$$

As $[m]$ and $[k]$ are symmetric, they remain as they are even after the transpose.

$\omega_n \neq \omega_m$ which is generally true except in some degenerate cases where $\omega_n = \omega_m$, i.e., certain frequencies of different modes could be equal to one another. In such cases even IS 1893-2016 has a procedure to be followed.

Because $\omega_n \neq \omega_m$, $\omega_n - \omega_m \neq 0$

$$\therefore \{\bar{x}\}_n^T [m] \{\bar{x}\}_m = 0 ....(c)$$

Similarly, $\{\bar{x}\}_n^T [k] \{\bar{x}\}_m = 0$. The first one is called orthogonality condition with respect to mass [m], and second one is orthogonality with respect to stiffness [k].

Equation (c) is the well-known orthogonal condition which is very useful to uncouple the coupled differential equation of motion in M.D.F After uncoupling the equations, of motion only one proceeds with the summation of modes, either with SRSS or CQC as mentioned in the IS 1893-16.

Shown in Fig. 7.5 is a typical building frame consisting of three storeys. The building frame consists of three floors. The first step in any vibration analysis is to identify the degrees of freedom. Let us demonstrate the analysis of a shear building.

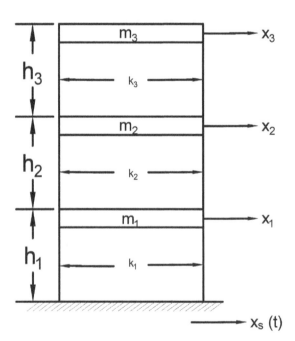

FIGURE 7.5: Typical three-storey building

(1) The masses are lumped at the floor levels only. Even the column masses taken equal to 50% below or above a floor are assumed to be concentrated at the floor level. It is quite a reasonable assumption because in all civil engineering residential and commercial buildings most of the load, dead load and live load are at the floor levels.

(2) The columns are assumed not to undergo rotations at the floor level. The above assumptions are quite valid, because in all tall reinforced concrete or steel buildings, the slabs are rigid in their own plane. Therefore, the columns are constrained from undergoing any rotation.

The assumptions make the analysis simpler, first, because the degrees of freedom to be identified become much less. Degrees of freedom have to be defined only at the mass points in a discrete system, and therefore as the masses are at the floor level only, the degrees of freedom, only lateral (horizontal) (as the earthquake force in lateral direction only) and are assumed at the floor levels. As a consequence, there will be as many degrees as the number of floors. As the degrees of freedom are the minimum required, it will be very computationally economical.

Displacement pattern of a shear building subjected to a lateral force such as an earthquake could be imagined to resemble that of a loaf of bread kept vertically with slices one above the other. The slice will slide forward in the direction of the lateral force when disturbed. For example, when the loaf is kept on the dining table and the table is horizontally moved for some reason.

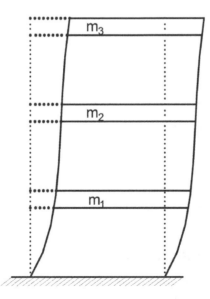

FIGURE 7.6a: Deformed shape of building (note: The rotations at the column joints at the base will be zero if the joint is fixed and it will be not zero if it is hinged)

The floors in the building have displaced similar to slices of bread kept vertically and distributed horizontally.

The horizontal stiffness of each floor of the above building is easily evaluated as $2 \times \frac{EI}{h^3}$, where $\frac{12EI}{h^3}$ is stiffness of each column, and there are two columns in each floor. It is similar to a beam fixed at two ends with the one support yielding by a unit distance.

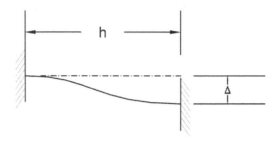

FIGURE 7.6b: A member of span h is subjected to support displacement Δ

Because there are only 3 DOF, the stiffness matrix is 3 and because there are only three masses, each one with one degree of freedom, the mass matrix is a 3×3 diagonal matrix.

$$\begin{bmatrix} m_1 & 0 & 0 \\ 0 & m_2 & 0 \\ 0 & 0 & m_3 \end{bmatrix}$$

The stiffness matrix is

$$\begin{bmatrix} k_{11} & k_{12} & k_{13} \\ k_{21} & k_{22} & k_{23} \\ k_{31} & k_{32} & k_{33} \end{bmatrix}$$

Similarly the damping matrix is

$$\begin{bmatrix} c_{11} & c_{12} & c_{13} \\ c_{21} & c_{22} & c_{23} \\ c_{31} & c_{32} & c_{33} \end{bmatrix}$$

The differential equation of motion is therefore coupled in stiffness and not in mass.

7.6 CAPACITY SPECTRUM

The next step in the improvement in the response spectrum method is the capacity spectrum approach. The capacity spectrum is a plot which gives both capacity of the structure and demand posed on it while resisting an earthquake.

An early paper on capacity spectrum method for determining the demand displacement is by Sigmund A. Freeman (ACI, 1994). The capacity spectrum method was first used to assess seismic vulnerability of buildings (Freeman, 1975) in a pilot project of the Puget sound naval shipyard. The capacity spectrum is a juxtaposition of pushover curve on the response spectrum. Pushover curve is obtained obviously by pushover analysis. Pushover analysis is a static non-linear analysis, which determines the lateral force or base shear required to cause yielding of the entire frame or in other words global yielding. It is obtained by applying a chosen lateral load either uniformly distributed at every storey level or even at one point which is the topmost storey. The load distributed along the height of the frame is generally chosen according to a mode shape and obviously the first mode. The lateral load is gradually increased and the moments at various joints, maybe girder end or column end, are monitored. As and when a particular moment reaches the yield value, a hinge is introduced. The yield moment is pre-determined based on the section properties, like area and moment of inertia of cross section, the yield strength of the material in the case of steel section or breadth and depth of cross section and reinforcement details and grade of concrete in the case of reinforced concrete. Thus, we obtain the yield moment M_y in steel member or $M_{u,lim}$, in the case of reinforced concrete section. As mentioned above once we note that a particular joint has reached M_y or $M_{u,lim}$, a hinge is introduced there and the stiffness matrix of the frame is revised with the hinge thus formed. At the previous load when the hinge formed, the lateral load on the frame, which is also the base shear is noted with the top storey lateral displacement. Thus, we have for one point in the pushover plot of lateral force/base shear vs. lateral top displacement. Similarly let us continue increasing the load further and analyze the frame with the revised stiffness matrix (with one hinge). As the frame has become relatively flexible with the hinge, the further pushover plot will become relatively flatter. Subsequently there will be another joint which will attain M_y or $M_{u,lim}$ as the case is. Then we will introduce a hinge there after noting the lateral load and the corresponding lateral top storey displacement. Thus, we keep on pushing the frame (therefore it is called pushover) by monotonically increasing the load till in the frame sufficient number of hinges are formed as to cause collapse. It requires n+1 hinges for a frame to form a collapse mechanism, where n is the degree of static indeterminacy. A plot of lateral force/base shear vs. lateral top storey displacement looking like a typical non-linear load-displacement plot is shown in Fig. 7.7.

Thus, we have determined the capacity curve of the frame as it yields under lateral force (earthquake force) till complete failure. Now we need to know the demand placed on the frame from a chosen earthquake for which the frame is designed. Obviously, the demand can be obtained from a response spectrum. Response spectrum is always expressed as a plot of maximum acceleration of a single degree of freedom system over various periods for a given earthquake. Because the pushover curve is a plot, force vs. max. top storey displacement, it is better to express the response spectrum also as such. Therefore, the response spectrum is plotted as max. acceleration vs. displacement and thus it is called as ADRS plot or in other words acceleration deformation response spectrum plot.

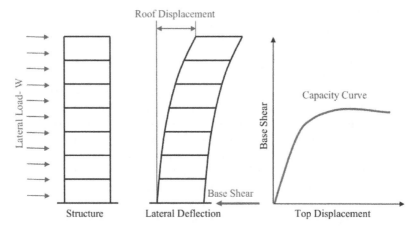

FIGURE 7.7: The basic concept of pushover

The frame whose pushover plot we thus obtained is also converted to an equivalent single degree of freedom system. The top storey displacement and base shear are converted to spectral displacement (S_d) and spectral acceleration (S_a), respectively, by the use of modal participation factors and effective modal masses determined for the first mode by usual procedure.

Thus, the expressions as proposed by Freeman, followed by Fajfar in their papers in 1998 and 1999 respectively. The two different plots are juxtaposed. The point of intersection gives the displacement at the corresponding acceleration demand. But the point here we are interested in the inelastic displacement or the displacement ductility required of the structure at the corresponding displacement ductility of the frame because the pushover curve is an inelastic curve. Therefore, it will be appropriate to juxtapose the inelastic response spectra or the pushover or covert the elastic response spectra to an approximate inelastic spectrum by the help of damping, which means juxtaposing damped response spectra on the pushover curve. The details can be obtained from the paper by Freeman and Fajfar.

In this method, the first step is the design of the structure and calculating the area of reinforcement required in structural element by seismic analysis procedure and then the pushover analysis is done.

The pushover curve obtained, i.e., V_b (base shear) vs. displacement curve is converted into corresponding S_a and S_d values on the capacity spectrum by equation (Fig. 7.8).

$$S_a = \frac{V_b/W}{M_k/M} g$$

$$S_d = \frac{\Delta_{rooftop}}{P_k \bar{x}_{rooftop}}$$

Earthquake Resistant Design

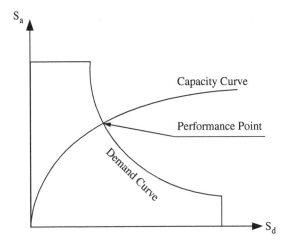

FIGURE 7.8: Performance point

where M_k = modal mass.
P_k = mode participation factor, and $\bar{x}_{rooftop}$ = modal amplitude at top floor for 1st mode.
M and W = total mass and weight of the structure.

The capacity and demand spectrum are superimposed and the performance (intersection) point is noted. This point shows the max. displacement the building will undergo during an earthquake before failure.

The studies have found that this method has a certain limitation when used for higher damped spectra and effective natural period corresponding to a performance point. Previous research showed that effective period at intersection point may not represent the true condition of the frame and also to non-linear dynamic response at higher target displacement. To overcome this limitation, several researchers have suggested the use of inelastic response spectra in place of elastic response spectra as demand spectrum, as it shows better estimates.

To convert the elastic spectra to inelastic response spectra following equations can be used as.

$$S_a = \frac{S_{ae}}{R_\mu}$$

$$S_d = \frac{\mu S_{de}}{R_\mu} = \frac{\mu T^2 S_{ae}}{R_\mu 4\pi^2} = \frac{\mu T^2 S_a}{4\pi^2}$$

Where μ = ductility factor $\frac{max.\ displacement}{yield\ displacement}$ and R_μ= reduction factor due to ductility.

$$R_\mu = \frac{(\mu - 1)T}{T_c} + 1, T < T_c$$

$R_\mu = \mu, T \geq T_c$

Here, T_c =characteristic period of the ground motion.

Capacity Spectrum Manual from SAP2000:

The pushover analysis is done on a G+2 structure of Example 3 of Chapter 6 using the SAP2000 software. In the model, beams and columns were modelled and default hinges properties were assigned. The diaphragm properties are assigned to slabs in every storey to establish lateral activity of beams. The analysis was carried out in both the principal directions but the results are shown only for x-direction because the structure is symmetrical. Fig. 7.9 shows the capacity spectrum plot obtained from the software as per ATC-40. The base shear and top displacement corresponding to performance point are 321.14 kN and 50.06 mm, respectively. At a performance point, the effective time period is 0.51 sec and effective damping is 17.6%. Fig. 7.10 shows

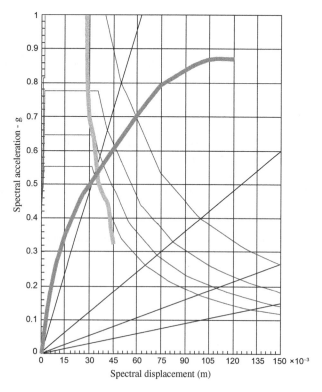

FIGURE 7.9: ADRS plot as per ATC-40

Earthquake Resistant Design

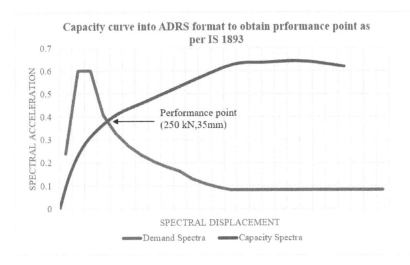

FIGURE 7.10: ADRS plot as per IS 1893:2016

the capacity spectrum obtained as per IS1893:2016. The corresponding base shear and top displacement are approximately 250 kN and 35 mm, respectively, as per manual calculations represented in the ADRS plot in Fig. 7.10, which was obtained using the procedure mentioned above. The ductility was found to be 8.37.

This failure mechanism is observed by the formation of hinges. Hinges are formed in a frame model when a node reaches its ultimate strength. Hinges are formed at the end parts of frame elements. These cracks are seen at a distance away from the joint and is the point where hinges are inserted in the frame components during computational analysis. The various types of hinges are axial, shear and flexural hinges. The shear and flexural hinges are assigned in beams as well as in columns as shown in Fig. 7.11. The axial hinges are assigned to a column and struts (modelled in place of walls to simulate its cracks).

Fig. 7.12 shows a moment-rotation of a typical beam, where the AB region represents the linear elastic range, A is an unloaded state and point B is the effective yield point. The slope BC represents reduced stiffness. Decrease in load resistance is seen in slope CD and from D to E it shows reduced resistance followed by F where component strength becomes zero.

The non-linear states of hinge within its ductile range (slope BC) are interpreted as immediate occupancy (IO), life safety (LS) and collapse prevention (CP). These points are usually obtained by dividing BC into four parts and are labelled as shown in Fig. 7.12. These levels in Fig. 7.12 are three performance levels of the performance-based design process. These levels represent performance of the frame under earth-

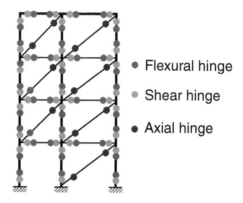

FIGURE 7.11: Hinge in different frame elements

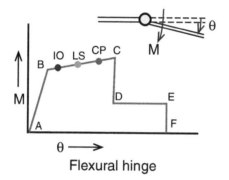

FIGURE 7.12: Moment rotation of typical beam

Table 7.1 Indication of performance levels

Levels	Indication
IO	Nominal damage to resisting elements and negligible damage to non-resisting elements
LS	Significant damage to both resisting & non-resisting element, repair work needs to be done before occupancy
CP	Failure of both resisting and non-resisting elements, permanent tilt in structure, most cases structure needs to be re-built

Earthquake Resistant Design

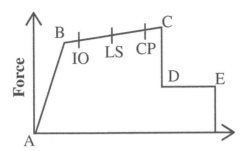

FIGURE 7.13: Performance levels

quake excitation. The capacity of the structure is shown by the capacity curve obtained from pushover analysis as shown in Fig. 7.7.

Table 7.2 gives the number of hinges formed at different levels in every step. The performance point of the structure lies between steps 4 and 5 in the table. If step 5 is taken as the value of performance point the structure performance lies in A-IO range as 28 hinges are formed in A-B range and 12 in B-IO range. Similarly, the performance point obtained using demand spectra of IS1893:2016 and from the same table the structure performance lies in A-IO range. Referring to Table 7.1 which shows indication of different levels, it can be said that structure is seismically safe (Fig. 7.13).

Table 7.2 Number of hinges

Step	Displacement mm	Base Shear kN	A-B	B-IO	IO-LS	LS-CP	CP-C	C-D	D-E	Beyond E	Total
0	0	0	40	0	0	0	0	0	0	0	40
1	5.59	94.019	38	2	0	0	0	0	0	0	40
2	12.71	173.03	34	6	0	0	0	0	0	0	40
3	21.49	231.15	30	10	0	0	0	0	0	0	40
4	36	290.39	28	12	0	0	0	0	0	0	40
5	54.	329.7	28	12	0	0	0	0	0	0	40
6	72	369.19	28	6	6	0	0	0	0	0	40
7	90	408.6	28	0	12	0	0	0	0	0	40
8	111	447.75	22	6	12	0	0	0	0	0	40
9	129	452.67	22	6	12	0	0	0	0	0	40
10	144.45	456.82	22	4	12	0	0	2	0	0	40
11	148.69	457.34	22	4	10	0	0	4	0	0	40
12	154.77	456.10	22	4	8	0	0	6	0	0	40
13	160.76	453.90	22	2	8	0	0	8	0	0	40
14	178.76	440.78	22	2	8	0	0	8	0	0	40
15	180	439.88	22	2	8	0	0	8	0	0	40

8 Inelastic Vibration Absorber Subjected to Earthquake Ground Motion

In the following Chapters 8, 9 and 10, the notations are different from the previous Chapters 1-7.

8.1 INTRODUCTION

In this and the following Chapters 9 and 10, work done by the author for his doctoral dissertation under the guidance of Prof. K.S. Jagdish, former professor in the Dept. of Civil Engineering, IISc, from 1973 to 1977 has been presented briefly.

In this chapter, we are discussing inelastic bilinear hysteretic vibration absorbers for earthquake ground motions. During 1970s, the first soft storey concept was in practice. Using that concept, several buildings were constructed to resist earthquakes. The concept was that most of the basements which are meant for car parking are very flexible because of no infill walls. It was found during earthquakes that a basement storey which is relatively more flexible compared to the storeys above, deflected more and in the event yielded at a number of points thus absorbing the energy of the earthquake.

The upper storeys remained intact although they drifted drastically along with the basement storeys. Apparently, there were no relative distortions and thus no damage. People in upper floors were safe. However, the building could not be reused because the basement itself was severely damaged and had severely swayed at one end. Thus, the building became a single-use building. The concept did not continue further on account of the above reason.

Therefore, it was suggested by the guide of the author, Prof. K.S. Jagdish, to think of a top soft storey which could also be called as expendable top storey, for a building to absorb the earthquake energy. The expendable top storey could be an immediate upper floor of a two-storey building or could be a topmost storey like a penthouse of any tall building. If such a top expendable storey is designed with not only the lowest lateral stiffness, but also lowest lateral yield lateral acceleration compared to those of the other lower storeys, then it could be called inelastic vibration absorber. Inelastic because it yields much before and more than lower storeys, thus, absorbing a lot of energy due to earthquake through hysteretic cycles. In the present example, such an inelastic vibration absorber or expendable top storey has been fitted to a one-storey building. The responses of the main storey have been obtained for different real earthquakes which occurred in the past. From the study, the optimum ratio of masses (ratio of the absorber mass to the main mass), ratio of stiffnesses and also yield accelerations and contour plots have been obtained.

Such an example as the above expendable storey is a classic example of a top storey being a soft as well as weak storey. In fact it is interesting to note such a concept is very much valued in IS 1893:2016. It is gratifying to note that the concept proposed by Prof. K.S. Jagdish and the author four decades ago has now come into IS code of practice and is taken more seriously.

Interestingly, one can observe failures of buildings during past earthquakes. The author has observed in detail the failures of two-storey structures during the Bhuj earthquake, although unfortunately those photographs are not available in this book. In those failures it has been observed that buildings had their ground floor completely collapse while the upper floor was fully intact and was just sitting on the floor. A similar two-storey buildings where the top floor completely collapsed, the ground floor was totally undamaged, which was an evidence of how the two-storey buildings behave when one of them is a weak storey. In the recent IS 1893:2016, a similar concept has been mentioned as a specification, namely soft and weak storeys are allowed only in the upper levels, or in other words a lower storey should not have lateral stiffness lower than that of the upper one while vice versa is preferred. In a similar way a lower storey should not be weaker than the upper one, while the vice versa is allowed.

8.2 THE LINEAR ELASTIC VIBRATION ABSORBER

Here, m_1 and k_1 are the mass and stiffness of a single degree freedom system subjected to a sinusoidal force $\overline{F}_1 sin\omega t$. Another mass m_a called absorber mass is connected to the main mass m_1 through the spring k_a after the absorber mass is connected to the system becomes a two degree of freedom system and it can be designed to have a zero displacement of mass m_1.

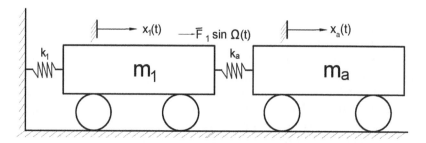

FIGURE 8.1a: Linear elastic vibration absorber attached to the main mass

Inelastic Vibration Absorber Subjected to Earthquake Ground Motion

Now writing the two equations of motion for the two degree freedom system,

$$m_1\ddot{x}_1 + (k_1 + k_a)x_1 - k_a x_a = \bar{F}_1 \sin\Omega t$$

$$m_2\ddot{x}_a - k_a x_1 + k_a x_a = 0$$

Let the steady-state solution be

$$x_1(t) = \bar{x}_1 \sin\Omega t, \quad x_a(t) = \bar{x}_a \sin\Omega t$$

$$\begin{bmatrix} k_1 + k_a - \Omega^2 m_1 & -k_a \\ -k_a & k_a - \Omega^2 m_a \end{bmatrix} \begin{bmatrix} \bar{x}_1 \\ \bar{x}_a \end{bmatrix} = \begin{bmatrix} \bar{F}_1 \\ 0 \end{bmatrix}$$

$$\bar{x}_1 = \frac{(k_a - \Omega^2 m_a)\bar{F}_1}{(k_1 + k_a - \Omega^2 m_1)(k_a - \Omega^2 m_a) - k_a^2}$$

$$\bar{x}_2 = \frac{k_a \bar{F}_1}{(k_1 + k_a - \Omega^2 m_1)(k_a - \Omega^2 m_a) - k_a^2}$$

Here,

$\omega_1 = \sqrt{\frac{k_1}{m_1}}$ is the natural frequency of the single degree freedom system.

$\omega_a = \sqrt{\frac{k_a}{m_a}}$ is the natural frequency of the absorber system.

$x_{st} = \frac{\bar{F}_1}{k_1}$ is the static deflection of the single degree freedom system.

$m_r = \frac{m_a}{m_1}$ is the ratio of the absorber mass to the main mass.

Further simplifying,

$$\bar{x}_1 = \frac{[1 - (\Omega/\omega_a)^2]x_{st}}{[1 + m_r(\omega_a/\omega)^2 - (\Omega/\omega)^2][1 - (\Omega/\omega_a)^2] - m_r(\omega_a/\omega)^2}$$

$$\bar{x}_a = \frac{x_{st}}{[1 + m_r(\omega_a/\omega)^2 - (\Omega/\omega)^2][1 - (\Omega/\omega_a)^2] - m_r(\omega_a/\omega)^2}$$

It can be easily seen that the above two expressions for the amplitudes, the amplitude of the main mass $\bar{x}_1 = 0$. When, $\Omega = \omega_a$.

The amplitude of the absorber mass is

$$\bar{x}_a = -(\frac{\omega}{\omega_a})^2 \frac{x_{st}}{m_r} = -\frac{\bar{F}_1}{k_a}$$

$$\therefore x_a(t) = \bar{x}_a \sin\Omega t = -\frac{\bar{F}_1}{k_a}\sin\Omega t$$

Force in the absorber spring is

$$k_a x_a(t) = -F_1 \sin\Omega t$$

which means the entire force is absorbed by the absorber in the spring-mass system and the main mass does not attract any force and thus will be at rest.

That is the beauty of linear vibration absorber. For example, when we come across a problem where a simple beam is carrying a motor. The rotor of the motor is rotating at N rpm. The beam is seen to vibrate very severely with large amplitude. We are supposed to stop the vibration of the beam. Then we take another spring mass system of mass m_a and spring stiffness k_a, such that $\omega_a = \sqrt{\frac{k_a}{m_a}}$ is equal to the forcing frequency $\Omega = \frac{2\pi N}{60}$ rad/sec. After we attach the absorber system, we will observe that the beam will come to rest while the absorber spring mass system will vibrate with the force $\overline{F}_1 \sin\Omega t$ or $\overline{F}_1 \sin\frac{2\pi N}{60}$.

Different varieties of absorbers have been designed in the past, for different kinds of inputs. Even for earthquake ground motion, linear absorbers have been designed. The reader can see the literature carried under inelastic vibration absorbers in the next chapter.

The principle of the vibration absorber has been studied extensively in literature and has found varied applications. A majority of these studies has considered vibration absorption in linear systems for sinusoidal inputs. However, some attempts have been made to extend the area of applicability of the absorber concept to include random inputs and non-linear effects.

The frequency response spectrum of the tuned absorber with optimum damping shows that the absorber action is present over a bandwidth in the neighborhood of the tuned frequency. This suggests that even for random inputs with the power spectrum spread over this bandwidth, the absorber effect may be realized. The response of the linear absorber subjected to stationary random inputs has been studied by Crandall and Mark (1963), Curtis and Boykin (1961) and Vasudeva Rao and Jagadish (1970). These studies have demonstrated the feasibility of using linear vibration absorber for random excitations. Many attempts have been made to investigate absorbers having various non-linear springs, with a view to improve the absorber action. Roberson (1952) used the well-known method of Duffing for a system with a cubic absorber spring. Pipes (1953) applied the same method to consider an absorber spring whose restoring force followed a hyperbolic sine variation. Arnold (1955) applied the Ritz averaging technique to an absorber with cubic spring and used a one-term approximation for the solution. Carter and Liu (1961) studied an absorber system where both the springs were non-linear. Bauer (1966) determined the magnitude of combination tones in an absorber system excited by two forces of different frequencies. In all the above cases, the exciting force was sinusoidal and it was found that a non-linear absorber behaved more effectively than a linear one.

Behavior of a variety of absorbers for earthquake ground motion has also been studied previously. Chandrashekaran and Gupta (1966) investigated a system wherein the main mass was connected by a number of vibration absorbers in parallel. They found that this was effective for strong ground motion type excitations and the reduction in the response of the main mass was found to be of the order of 60 to 70% of that without the absorber. Gupta and Chandrashekaran also studied the dry friction type of absorbers and showed that these could be used effectively for earthquake-type ground motion. The reduction in response of the main mass was 35% of that without the absorber. The ground motions considered were ElCentro earthquake, May 1940, N-S component and Taft record, July 1952 $S21°N$.

Wirshing and Yao (1970) carried out a statistical study of some design concepts in earthquake engineering. They studied an absorber system subjected to an ensemble of random samples whose intensity was matched to be very nearly equal to that of N-S Component of the El Centro, May 1940 record. It was found that an absorber system could be as effective as to reduce the mean response of the main system by about 30 to 40%.

Ohno, Watari and Sano (1977) analytically arrived at optimum tuning for linear dynamic absorber subjected to earthquake ground motion.

8.3 THE HYSTERIC VIBRATION ABSORBER

It is now well established that most of the structures undergo yielding during strong ground motions. This means that structures undergo non-linear hysteric deformations during earthquakes. In this chapter, the possibilities of absorber behavior have been explored for yielding structures subjected to earthquake ground motions. A two-storey, bilinear hysteric structure has been chosen for the purpose of this study. The objective of the investigation is to find the circumstances under which the top storey of the structure could absorb a major portion of the energy input, thus reducing the response levels of the lower storey. If such a behavior is feasible, one can conceive of a structure whose top storey is permitted and designed to undergo large inelastic deformations, while reducing damage in the lower storey. Such a design approach may well be termed as the expendable top storey concept. It may well be remarked that such a concept juxtaposes the often mentioned soft first-storey concept. The soft first-storey approach has been commented upon widely for the questions it poses about the stability and safety of the design. It looks as though the expendable top storey concept is more favorably placed regarding the overall stability and safety of the structure. The top storey would then behave like a mechanical fuse undergoing large deformations. However, the problem of designing the top storey to withstand the large ductility demand placed on it needs to be looked into. The fact that the absorber system has a lower mass compared to that of the main structure probably favors the practical design of the top storey.

8.4 STRUCTURAL MODEL AND THE EQUATIONS OF MOTION

Fig. 8.1 shows a two-storey hysteric structure. Its base is excited by a horizontal acceleration of $\ddot{X}_0(t)$ due to an earthquake. The lower-storey mass is designated by m_1 and the absorber mass by m_2. The two springs follow a bilinear hysteric, force-displacement relationship as shown in Fig. 8.2. K_1 and K_2 are the stiffnesses of the main mass and the absorber spring in the elastic regime.

After yielding, the springs follow the lines with 10% of the slope in the elastic region. It is also assumed that both the storeys have viscous damping mechanisms represented by the dash-pot coefficients C_1, and C_2, $X_1(t)$, $X_2(t)$, the absolute displacements of the two masses are the generalized coordinates of the system.

The equations of motion in the non-dimensional form may be written as

$$Z_1'' + 2S_1 Z_1' + p(Z_1) - 2\mu S_2 F Y Z_2' - \mu Y F^2 p(Z_2) = -\frac{\ddot{X}_0(t)}{q_Y} \tag{8.1}$$

$$Z_1'' + Y Z_2'' + 2S_2 F Y Z_2' + Y F^2 p(Z_2) = -\frac{\ddot{X}_0(t)}{q_Y} \tag{8.2}$$

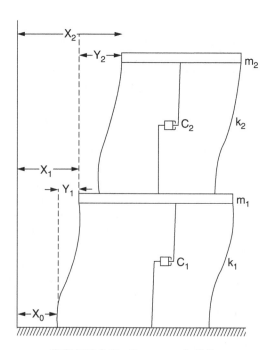

FIGURE 8.1b: Two-storey building

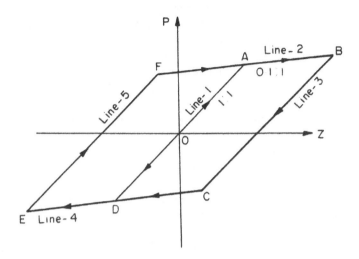

FIGURE 8.2: Force-displacement relationship

The primes indicate derivatives with respect to τ, a non-dimensional time unit. Here,

$Z_1 = \frac{X_1 - X_0}{Y_1}$, non-dimensional displacement of the bottom-storey relative to the base.

$Z_2 = \frac{X_2 - X_1}{Y_{II}}$, non-dimensional displacement of the top-storey relative to the bottom storey.

Y_1 and Y_{II}, relative yield displacements of the bottom and top storey, respectively.

$p(Z_1)$ and $p(Z_2)$, non-dimensional resistance forces of the bottom and top storey, respectively.

$S_1 = C_1/2\sqrt{K_1 m_1}$
$S_2 = C_2/2\sqrt{K_2 m_2}$
C_1 and C_2, damping constants of the bottom and top storey, respectively.
$\tau = \omega_1 t$
$\omega_1 = \sqrt{K_1/m_1}$
$\omega_2 = \sqrt{K_2/m_2}$
$F = \frac{\omega_2}{\omega_1}$ — ratio of frequencies
$Y = \frac{Y_{II}}{Y_1}$ — ratio of yield displacements
$\mu a = \frac{m_2}{m_1}$ — ratio of masses
q_Y = acceleration required to cause yielding of the bottom storey expressed as a fraction of the acceleration due to gravity, g.

8.5 NUMERICAL STUDIES

The equations of motion (8.1) and (8.2) were solved numerically using the linear acceleration method on IBM 360/44 Computer. A step size of 1/70 of the period of the lower storey has been chosen for the integration. This step size was found to give satisfactory results. The Taft 1952, S69°E earthquake accelerogram has been chosen as the input for all the systems studied here.

The maximum excursion of Z_1 or Z_2 in the inelastic region during a cycle is defined as the ductility factor and is denoted by D_1 and D_2 for the bottom and top-storey, respectively. The maximum value of D_1 and D_2 attained over the duration of an earthquake is denoted by $\overline{D_1}$ and $\overline{D_2}$, respectively. The behavior of $\overline{D_1}$ and $\overline{D_2}$ has been studied in detail for various cases. The other features of response observed are

(1) The dominant period of vibration
(2) The energy dissipated due to hysteresis.

A variety of two-storey systems is considered for the Taft S69°E, 1952 earthquake accelerogram. The influences of the following parameters are examined:

(a) Mass ratio μ, (b) frequency ratio F, (c) yield displacement ratio Y, (d) yield strength of the bottom storey q_Y, (e) damping coefficients S_1 and S_2, (f) linear frequency of the bottom storey. Table 8.1 gives the range of values chosen for each of the above parameters.

TABLE 8.1: Input details

No.	Parameter	Range of numerical values assigned
1	Mass ratio (μ)	0.1, 0.2, 0.3 and 1.0
2	Frequency ratio (F)	0.2 to 1.2
3	Yield displacement ratio (Y)	0.2 to 1.2
4	Yield strength of the bottom storey (q_Y)	0.1 g, 0.2 g, and 0.3 g
5	Damping coefficients (S_1 and S_2)	1.0% and 5.0%
6	Period of the lower-storey (T)	0.42 sec and 1.0 sec

8.6 ANALYSIS OF RESULTS

The maximum deflection ductilities $\overline{D_1}$ and $\overline{D_2}$ of the two storeys for various system parameters have been studied in some detail. The results have been presented in the form of contours of constant maximum ductility $\overline{D_1}$ and the histories of dynamic displacements for typical cases. Figs. 8.3 to 8.9 show maximum ductility $\overline{D_1}$ contours for the lower storey. The mass ratio, the viscous damping factor and the yield acceleration of the lower storey are held constant for each figure. The contours indicate the influence of the frequency and yield displacement ratios on the maximum ductility response of the lower storey. The maximum ductility response of the structure with the absorber removed is indicated on each graph by the symbol $\overline{D_0}$.

Figures 8.3 to 8.5 show the ductility contours for the lower storey for various mass ratios for $q_Y = 0.3g$ g being the acceleration due to gravity. The contours show that there are valleys in the maximum ductility surface where the maximum ductility values are significantly smaller than the corresponding single-storey result. In particular, Fig. 8.4 shows that for a mass ratio of 0.3, the maximum ductility can be less than 1-0, for frequency ratio of 0.9 and yield ratios between 0.6 and 1.0. The maximum response is thus reduced by more than 50% by the addition of the absorber system. It may also be noted that this vibration absorption is available over a wide range of parameters. For instance, referring back to Fig. 8.4, even if the frequency ratio is varied from 0.8 to 1.0 and the yield ratio varied between 0.5 to 1.1, the maximum ductility is not very different from 1.2. Similar results are obtained for mass ratios $\mu = 0.1, 0.2$ and 1.0 although the magnitude of vibration absorption is not as significant.

The region over which absorption is obtained also shifts with a change in the mass ratio. Similar contours for $q_Y = 0.2$ g and 0.1 g are presented in Figs. 8.6 and 8.7, respectively. The contours for $q_Y = 0.2$ g are again similar. The maximum ductility of the lower storey is again brought down by about 50% by the absorber action. However, the region over which this reduction is obtained is narrower for $q_Y = 0.2$ g compared to $q_Y = 0.3g$. It is interesting to note that the vibration absorber behavior is not appreciable when the yield acceleration is 0.1g. The maximum amplitude reduction is 25% and is available only at one point, viz., F = 1.0, Y = 1.2. Even in the immediate

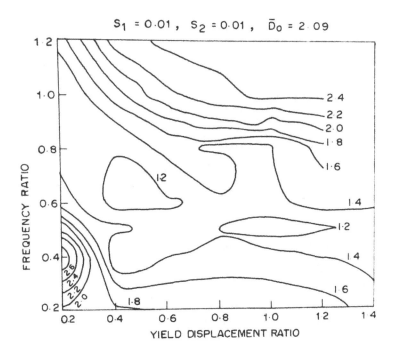

FIGURE 8.3: Maximum ductility contours, $q_Y = 0.3g$, $\mu = 1.0$

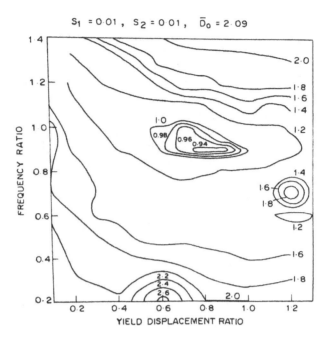

FIGURE 8.4: Maximum ductility contours, $q_Y = 0.3g$, $\mu = 0.3$

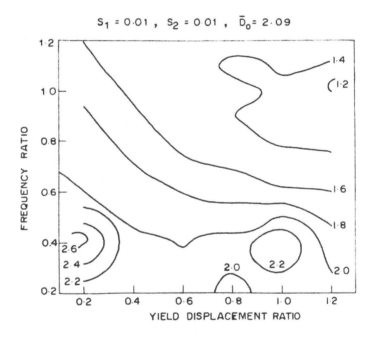

FIGURE 8.5: Maximum ductility contours, $q_Y = 0.3g$, $\mu = 0.1$

Inelastic Vibration Absorber Subjected to Earthquake Ground Motion

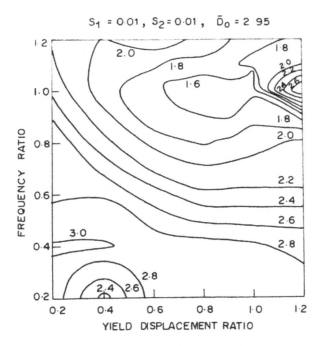

FIGURE 8.6: Maximum ductility contours, $q_Y = 0.3g$, $\mu = 0.3$

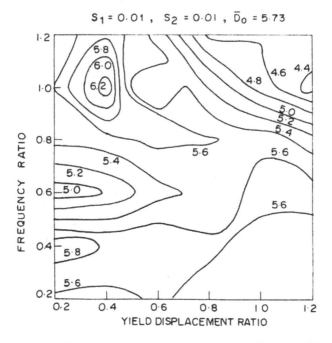

FIGURE 8.7: Maximum ductility contours, $q_Y = 0.1g$, $\mu = 0.3$

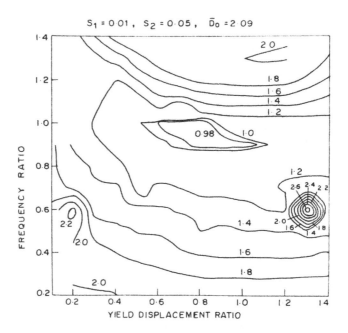

FIGURE 8.8: Maximum ductility contours, $q_Y = 0.3g$, $\mu = 0.3$

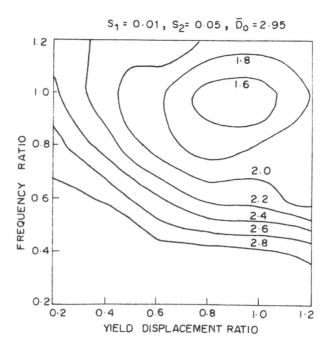

FIGURE 8.9: Maximum ductility contours, $q_Y = 0.2g$, $\mu = 0.3$

neighborhood of this point the reduction is decreased. The range over which at least 25% reduction is present is extremely narrow.

8.6.1 RESPONSE OF THE ABSORBER MASS

Figs. 8.10 and 8.11 compare the maximum ductility response of the top mass with that of the lower mass for typical system parameters. In general, the top mass experiences much larger ductilities than the lower mass. It is also seen that as the strength of the upper storey is reduced, its maximum displacement increases in proportion. This is completely in contrast with the response characteristics of the lower storey. The response curves of the lower storey dip and attain minimum values as the yield displacement ratio approaches unity and the frequency ratio is between 0.8 and 1.0.

It is instructive to compare these results with some of the known results of linear vibration absorber. In un-damped absorber subjected to sinusoidal input, a frequency

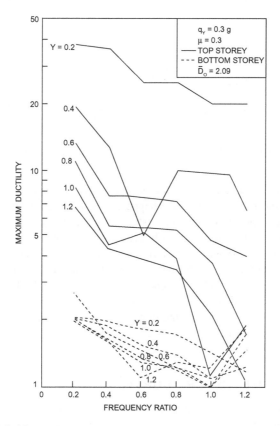

FIGURE 8.10: Maximum ductility vs. frequency ratio for, $S_1 0.01$, $S_2 = 0.01$

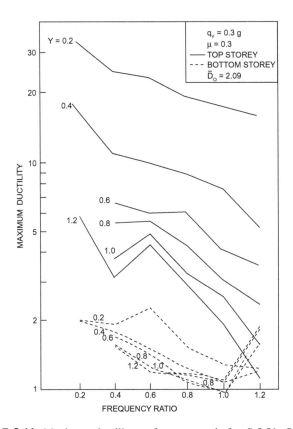

FIGURE 8.11: Maximum ductility vs. frequency ratio for, $S_1 0.01$, $S_2 = 0.05$

ratio of unity will yield a best absorber system. The frequency ratio in the case of most favorably tuned damped absorber happens to be around the value 0.9. Maximum reduction in the response of the main system subjected to filtered Gaussian input is obtained by attaching an absorber with frequency ratios between 0.65 and 0.92. In linear systems, with earthquake type input, a frequency ratio of unity has been found to be most suitable for best performance as an absorber. This feature is repeated in hysteric absorber systems and a frequency ratio, F, in the neighborhood of unity gives good absorber action. It is interesting to note that, in spite of the non-linearities of the system due to inelastic action, the frequency ratio $F = 1$ has a strong influence in providing dynamic vibration absorption.

8.6.2 RESPONSE HISTORY CURVES

The response histories of the displacements in typical absorber systems are shown in Figs. 8.12 through 8.17. The response envelopes of the corresponding single storey structure (SDF) without the absorber are also shown superposed on the response of the lower storey. The figures give an insight into the mechanism of vibration absorption in yielding systems. Fig. 8.12 shows the response of a system with F = 1.0, Y = 0.8 and $\mu = 0.3$ for $q_Y = 0.3$ g, where good vibration absorption is noticed. It is clear that it is not merely the maximum response that is brought down by the vibration absorber, but also the vibration levels of the lower storey are reduced throughout, due to the presence of the absorber. It may also be noted that the dominant periods of vibration of the top and bottom storeys, denoted by T_t and T_B, respectively, are 0.51 and 0.41 sec. The two periods are fairly close. Figure 8.16 shows that when the yield ratio is very small, the top and bottom storeys can execute vibrations with different dominant periods, viz., $T_t = 3.12$ sec and $T_B = 0.45$ sec, respectively. This may be considered as typical of non-linear system response wherein the concept of normal modes is no longer relevant. The top mass vibrates at its own frequency and incidentally there is no absorber action for this case. Similar conclusions can be inferred on observing Figs. 8.12 through 8.17. The values of T_t and T_B are shown in the figures. The top and bottom storeys vibrate at frequencies which are close when the frequency ratio is near 1.0 and the yield ratio is not very small. In a sense, this result is in harmony with the linear vibration absorber theory, which requires close natural frequencies for best absorber action.

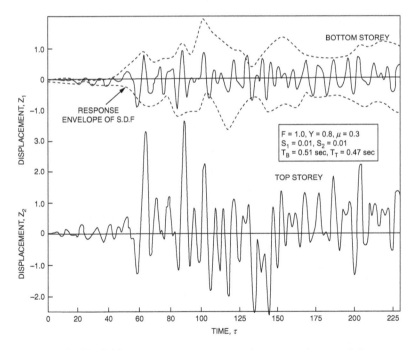

FIGURE 8.12: Displacement response - time curves for $q_Y = 0.3g$

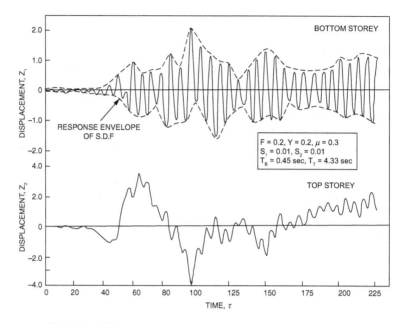

FIGURE 8.13: Displacement response - time curves for $q_Y = 0.3g$

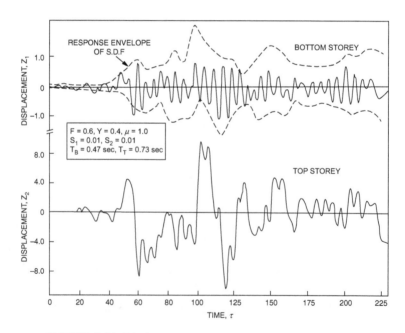

FIGURE 8.14: Displacement response - time curves for $q_Y = 0.3g$

Inelastic Vibration Absorber Subjected to Earthquake Ground Motion 123

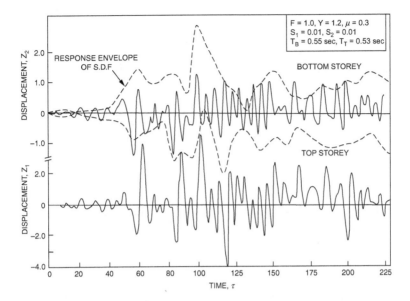

FIGURE 8.15: Displacement response - time curves for $q_Y = 0.2g$

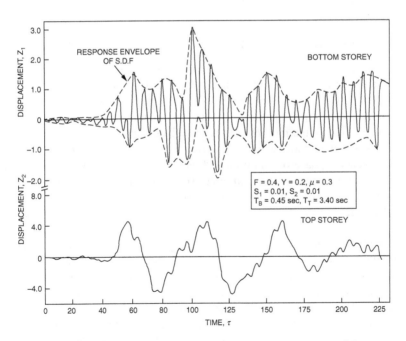

FIGURE 8.16: Displacement response - time curves for $q_Y = 0.2g$

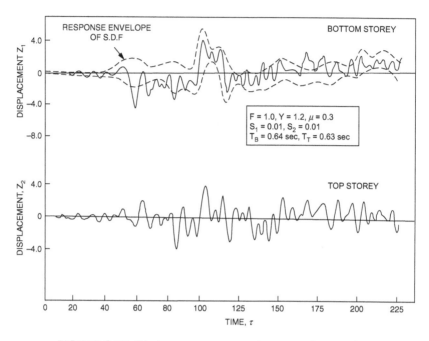

FIGURE 8.17: Displacement response - time curves for $q_Y = 0.1g$

8.6.3 HYSTERIC ENERGY DISSIPATION

Tables 8.2 through 8.5 display the ratios of the total hysteric energy absorbed by the bottom storey up to the end of the response H_B to that by the top storey H_T. It can be observed that in all the cases presented in the tables, the ratio H_B/H_T decreases as maximum ductility D_1 decreases. Very often, at the points where there is maximum amplitude reduction, the ratio H_B/H_T is zero. This implies that the response of the main mass is in the elastic region. The system shows no hysteric energy dissipation during the elastic phase of vibration and whatever energy dissipation takes place is accounted for by the viscous damping. At the same time, it is also clear that for values of F and Y for which the absorber action is not good the ratio H_B/H_T, is generally large.

Referring to Table 8.2, for the case $q_Y = 0.3$ g, F = 1.0, Y = 0.8, the ratio H_B/H_T, is zero and here the amplitude reduction due to the absorber is nearly 50%. From the time history for this case in Fig. 8.12, it is evident that both the storeys execute almost the same number of 3 cycles, the top mass entering the inelastic range very frequently. The closeness of the dominant periods of vibration, which is 0.54 sec for the bottom storey T_B and 0.47 sec for the top storey, T_T, explains the good absorber 3 action. For low values of F and Y such as F = 0.2, Y = 0.2 the absorber action is not present and the ratio H_B/H_T is 25-20. From Fig. 8.13 it can be seen that the dominant period of the bottom storey T_B is 0.46 sec and that of the top storey T_T

Table 8.2 Ratios of maximum ductility and hysteric energy dissipation, $q_Y = 0.3$ g T = 0.46 sec

No. (1)	Mass ratio, μ (2)	Viscous damping coefficients S_1 (3)	S_2 (4)	Frequency ratio, F (5)	Yield displacement ratio, Y (6)	$\overline{D}_1/\overline{D}_2$ (7)	$\overline{D}_1/\overline{D}_0$ (8)	Hysteretic energy dissipated by bottom storey, H_B / Hystreretic energy dissipated by top storey, H_T (9)
1	0.3	0.01	0.01	0.2	0.2	0.05	0.98	25.20
				0.2	0.8	0.18	0.96	62.50
				0.6	0.8	0.28	0.65	0.35
				0.8	0.8	0.23	0.56	0.025
				1.0	0.2	0.07	0.68	0.0
				1.0	0.6	0.22	0.48	0.0
				1.0	0.8	0.27	0.48	0.0
2	0.3	0.01	0.05	0.2	0.8	0.22	0.95	17.30
				0.4	0.8	0.29	0.75	1.15
				0.8	0.8	0.26	0.53	0.04
				1.0	0.6	0.24	0.47	0.0
				1.0	0.8	0.32	0.46	0.0
				1.0	1.0	0.42	0.52	0.0

(1)	(2)	(3)	(4)	(5)	(6)	(7)	(8)	(9)
3	0.2	0.01	0.01	0.2	0.2	0.05	0.99	40.33
				0.2	0.8	0.20	0.96	11.07
				0.6	0.6	0.21	0.74	19.47
				1.0	0.6	0.18	0.53	0.0
				1.0	0.8	0.25	0.48	0.0
4	0.1	0.01	0.01	0.2	0.6	0.16	1.00	41.30
				0.4	1.2	0.53	0.91	8.21
				0.8	0.2	0.07	0.90	35.35
				1.0	1.0	0.38	0.64	0.0
				1.2	1.2	0.89	0.84	0.06
5	1.0	0.01	0.01	0.2	0.6	0.14	0.86	5.08
				0.2	1.2	0.25	0.81	7.62
				0.6	0.4	0.11	0.54	0.0
				0.6	0.6	0.19	0.56	0.0
				0.6	1.0	0.32	0.65	0.0

$\overline{D}_1$ = Maximum ductility of bottom storey; $\overline{D}_2$ = Maximum ductility of top storey
$\overline{D}_0$ = Maximum ductility of the structure (SDF) with the absorber removed

is 4.33 sec; it appears, as though, the wide disparity in the periods of the bottom and top storeys is responsible for the unfavorable results. While both storeys show excursions beyond the yield level, excursions of the bottom storey outnumber those of the top H storey. Consequently, the cumulative area of the hysteresis loops traced by the top storey is very small compared to that of the bottom storey. The top storey response shows a superposition of a low frequency oscillation with period 4.353 sec and a high frequency oscillation with period 0.46 sec. The smaller period of vibration

Table 8.3 Ratios of maximum ductility and hysteric energy dissipation, $q_Y = 0.2$ g
T = 0.46 sec

No.	Mass ratio, μ	Viscous damping coefficients S_1	S_2	Frequency ratio, F	Yield displacement ratio, Y	$\overline{D}_1/\overline{D}_2$	$\overline{D}_1/\overline{D}_0$	Hysteretic energy dissipated by bottom storey, H_B / Hysteretic energy dissipated by top storey, H_T
(1)	(2)	(3)	(4)	(5)	(6)	(7)	(8)	(9)
1	0.3	0.01	0.01	0.2	0.8	0.21	0.98	11.58
				0.4	0.8	0.34	0.98	37.20
				0.8	0.8	0.27	0.60	0.20
				1.0	0.8	0.31	0.48	0.11
				1.2	0.2	0.08	0.73	0.0
2	0.3	0.01	0.05	0.2	0.2	0.06	0.99	50.00
				0.4	0.8	0.40	0.97	6.87
				0.6	1.0	0.42	0.71	1.10
				0.8	1.0	0.41	0.61	0.16
				1.2	1.0	0.91	0.65	0.0

(1)	(2)	(3)	(4)	(5)	(6)	(7)	(8)	(9)
3	0.2	0.01	0.01	0.2	1.2	0.30	0.99	33.08
				0.4	1.2	0.36	0.96	5.10
				0.6	1.2	0.47	0.80	2.29
				1.0	0.8	0.36	0.68	0.10
				1.0	1.2	0.55	0.58	0.17

$\overline{D}_1$ = Maximum ductility of bottom storey; $\overline{D}_2$ = Maximum ductility of top storey
$\overline{D}_0$ = Maximum ductility of the structure (SDF) with the absorber removed

Table 8.4 Ratios of maximum ductility and hysteric energy dissipation, $q_Y = 0.1$ g
T = 0.46 sec

No.	Mass ratio, μ	Viscous damping coefficients S_1	S_2	Frequency ratio, F	Yield displacement ratio Y	$\overline{D}_1/\overline{D}_2$	$\overline{D}_1/\overline{D}_0$	Hysteretic energy dissipated by bottom storey, H_B / Hysteretic energy dissipated by top storey, H_T
1	0.3	0.01	0.01	0.2	0.2	0.05	0.97	107.33
				0.4	1.2	0.51	0.95	7.77
				0.6	0.6	0.29	0.93	3.02
				1.0	0.6	0.54	0.76	0.53
				1.0	1.2	1.06	0.75	0.66

$\overline{D}_1$ = Maximum ductility of bottom storey; $\overline{D}_2$ = Maximum ductility of top storey
$\overline{D}_0$ = Maximum ductility of the structure (SDF) with the absorber removed

Inelastic Vibration Absorber Subjected to Earthquake Ground Motion

Table 8.5 Ratios of maximum ductility and hysteric energy dissipation, $q_Y = 0.3$ g
T = 0.46 sec

No.	Mass ratio, μ	Viscous damping coefficients S_1 S_2	Frequency ratio, F	Yield displacement ratio, Y	$\overline{D}_1/\overline{D}_2$	$\overline{D}_1/\overline{D}_0$	Hysteretic energy dissipated by bottom storey, H_B / Hysteretic energy dissipated by top storey, H_T
1	0.3	0.05 0.01	0.2	0.6	0.11	0.96	0.02
			0.2	1.2	0.22	0.95	0.25
			0.6	0.6	0.18	0.77	0.0006
			1.0	0.6	0.21	0.57	0.0
			1.0	0.8	0.28	0.57	0.0
			0.6	1.2	0.19	0.62	0.0

$\overline{D}_1$ = *Maximum ductility of bottom storey;* $\overline{D}_2$ = *Maximum ductility of top storey*
$\overline{D}_0$ = *Maximum ductility of the structure (SDF) with the absorber removed*

corresponds to the dominant response period of the bottom storey, but the amplitudes are very small. The maximum amplitude of the top storey is contributed by the large period oscillation. The small period oscillation is superposed on this in the form of clumping phenomena. Energy absorbed by the top storey is necessarily small because the number of cycles corresponding to the large amplitude oscillation is as few as three. The history curve shows an interesting example of non-linear response where the normal mode concept is not relevant.

For $q_Y = 0.2$ g (Table 8.3), mass ratio 0.3, at F = 1.0, Y = 0.8, the value of H_B/H_T is 0.11. The dominant periods of the bottom and top storeys are 0.55 sec and 0.53 sec, respectively. Again, here they are close. For F = 0.4, Y = 0.2 (Fig. 8.16), there are hardly three complete cycles for the total duration of the response with amplitude greater than 1·0 in the top storey. In contrast, the bottom storey executes more than 18 full cycles. The dominant period of the bottom storey, T_B is 0.45 sec and that of the top storey, T_T is 3.4 sec.

In the case of $q_Y = 0.1$g (Table 8.4), the magnitude of absorption is not very favorable. There are a few cases where the ratio H_B/H_T is less than 1.0. However, the reduction in amplitude of the bottom storey is not significant. Fig. 8.17 gives response-time curves for F = 1.0, Y = 1.2. In this case, the amplitude reduction is 25% which is best for $q_Y = 0.1$ g. Here also, as expected, the dominant periods of vibration of the bottom and top storeys are very close, viz. 0.64 sec and 0.63 sec, respectively. A ratio of H_B/H_T of 0.66 obviously indicates that top storey has dissipated more energy through hysteresis. Although this has caused some reduction of the bottom storey amplitude, it does not appear to be significant. Fig. 8.20 showing the cumulative hysteresis energy dissipation in the two storeys as a function of time τ reveals certain features of this response behavior. It is interesting to note that for F = 1.0, Y = 0.4, hysteresis energy dissipated by the bottom storey is very nearly equal to that dissipated by the single degree of freedom system (with the absorber absent).The ratio, H_B/H_T is equal to 2.64. There is no amplitude reduction due to the presence of the absorber. Instead, there is an increase of 10% of that of an SDF system in the amplitude of the lower storey. Further, the bulk of the hysteresis energy dissipation occurs between τ = 90 and τ = 120 due to a transient peak in the response (Fig. 8.17). However, when F = 1.0 and Y = 1.2, although the reduction in the bottom storey amplitude is only 25%, the energy dissipated by the bottom spring is about a third of the energy dissipated by a corresponding SDF system. It may also be noted that the sudden spurt of the energy dissipation present in the SDF system is no longer, sec in the system with the absorber. It may be remarked here that the maximum response amplitudes $\overline{D_1}$ do not give a complete picture of the absorber action. Although, the amplitude reduction does not appear to be satisfactory for q_Y = 0.1 g, reduction in the hysteresis energy dissipation is significant.

Some typical cumulative hysteresis energy dissipation curves also for q_Y = 0.2 g are presented in Figs. 8.18 and 8.19. It is interesting to note that, whore the absorber action is it not good (Fig. 5.19), the curves for the bottom storey and the SDF system practically follow each other. In such cases, the top storey is more or less ineffective. Fig. 8.18 shows an instance where the vibration absorption is good. Here, the bottom storey stops dissipating hysteresis energy at the middle of the ground motion duration and the cumulative energy dissipated by the top storey grows rapidly at this point of time. The total energy dissipated by the top storey, in this case, is practically equal to that of the SDF system. It may be observed, in general, that it is during the strong

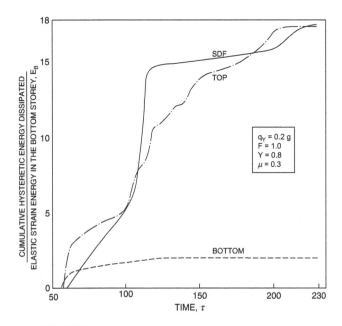

FIGURE 8.18: Cumulative hysteretic energy dissipated vs. time τ

Inelastic Vibration Absorber Subjected to Earthquake Ground Motion

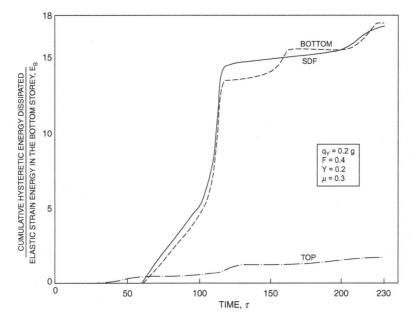

FIGURE 8.19: Cumulative hysteretic energy dissipated vs. time τ

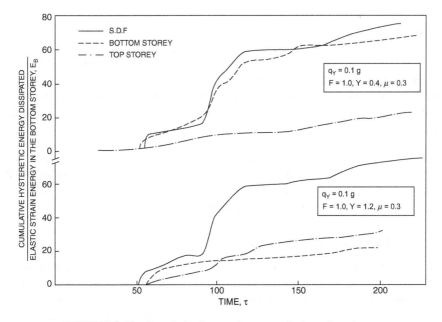

FIGURE 8.20: Cumulative hysteretic energy dissipated vs. time τ

motion phase of the earthquake the bulk of the hysteresis energy dissipation takes place, whether it be in the top storey or the bottom storey.

8.6.4 INFLUENCE OF VISCOUS DAMPING

In linear absorber systems subjected to filtered Gaussian input, an increase in viscous damping coefficient of the lower spring worsened the absorber action. At the same-time, an increase in the viscous damping up to 10% of the critical in the top storey improved the absorber effect. Any increase, beyond this level, increased the amplitude levels of the main mass.

In the present study, the viscous damping coefficient has been varied both in the top and bottom storeys independently.

Figs. 8.8 and 8.9 show the maximum ductility contours when the viscous damping coefficient S_2 in the top storey is 5%. A comparison of this with the contours for top storey damping S_2 of 1% is interesting. There is a close resemblance between the two sets of contour plots. Some of the local peaks and valleys have been smoothened out in the former case. Also the smoothening effect is very significant in Figs. 8.10 and 8.11, showing $\overline{D_2}$ vs. F relationship. With $S_2 = 0.05$, the amplitude level of the main mass is further brought down. It can also be observed that the amount of MI hysteretic energy absorption decreases with an increase in the viscous damping factor (Tables 8.2 and 8.3). This might indicate that the total energy dissipated is shared between hysteretic cycles and viscous damping.

Table 8.5 shows the effect of increasing S_1 to 0.05. There is a marked change, for the better, in the reduction of the ductility level of the main mass. For e.g., for $q_Y = 0.3$ g, ductility demand of the lower spring is only 42% of that of single degree-of-freedom system with $S_1 = 0.01$. It was found earlier that for the same case with $S_1 = 0.01$, the reduction was about 50%. This is very much in contrast with the behavior of linear absorbers. However, such improved absorber effect is not found with $q_Y = 0.1$g, for $S_1 = 0.05$.

Systems with linear elastic period of 1.0 sec remain in the elastic range with and without the absorber. For such cases, an increase in the top storey damping improved the situation which is in conformity with the linear absorber behavior. The results for such systems have been presented in Table 8.6.

8.6.5 MAXIMUM DUCTILITY RESPONSE SPECTRA

The spectra for a typical inelastic vibration absorber, showing the maximum ductility response of the two storeys, is shown in Fig. 8.21. The spectrum for the corresponding SDF system is also shown for comparison. It is interesting to note that the absorber action reduces the maximum response of the lower storey over periods ranging from 0.13 to 1.10 sec. Incidentally, this is the range of periods over which the SDF response

Table 8.6 Ratios of maximum ductility and hysteric energy dissipation, $q_Y = 0.3$ g T = 1.0 sec

No.	Mass ratio, μ	Viscous damping coefficients S_1	S_2	Frequency ratio F	Yield displacement ratio, Y	$\bar{D}_1 / \bar{D}_2$	$\bar{D}_1 / \bar{D}_0$	Hysteretic energy dissipated by bottom storey, H_B / Hysteretic energy dissipated by top storey, H_T
1	0.3	0.01	0.01	0.2	0.4	0.14	0.89	0.0 / 0.0
				0.2	0.8	0.34	0.89	0.0 / 0.0
				0.2	1.0	0.45	0.89	0.0 / 0.0
				0.4	0.2	0.07	0.80	0.0 / 0.0
				0.8	0.8	0.31	0.79	0.0 / 0.0
				0.6	0.2	0.07	0.77	0.0 / 0.0
2	0.3	0.01	0.05	0.2	0.4	0.15	0.85	0.0 / 0.0
				0.2	0.8	0.35	0.85	0.0 / 0.0
				0.2	1.0	0.45	0.85	0.0 / 0.0
				0.4	0.2	0.07	0.77	0.0 / 0.0
				0.4	0.8	0.31	0.74	0.0 / 0.0
				0.6	0.2	0.07	0.72	0.0 / 0.0

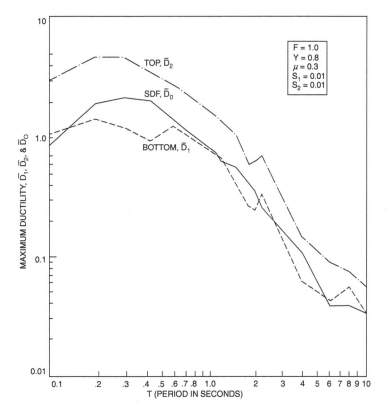

FIGURE 8.21: Maximum ductility response spectra for $q_Y = 0.3$ g

is maximum. For higher values of the linear period, the response levels are comparable. It may be emphasised that the SDF response levels are not serious for the longer periods. It is further interesting to note that the response of the lower storey does not become adverse at other periods.

The spectra shows that the absorber system can effectively reduce the amplitudes of the lower storey over that range of periods where the amplitude reduction is desirable from the designer's point of view.

8.7 CONCLUSIONS

Studies reported in this chapter have shown the possibility of designing inelastic two-storey structures, where the top storey behaves like a vibration absorber for horizontal ground motions. The numerical investigations made for the Taft 1952, S69°E component earthquake indicates that it is possible to achieve a 50% reduction in the ductility demand made on the lower storey. However, the ductility demand on the absorber storey would be very much larger than what may be normally expected. It has also been found for the particular earthquake ground motion considered, that the lower storey should have a moderately large yield acceleration of 0.3 g for good vibration absorption. The frequency ratio and the yield displacement ratio of the two storeys might have values between 0.8-1.0 for best results.

Although the maximum amplitude reduction achieved is only about 50%, the effect of vibration absorber can be seen more clearly in a study of the hysteric energy absorption. It may be seen that when the absorber is effective, it often dissipates about ten times the energy dissipated by the lower storey. It may be concluded that a major portion of the input energy is siphoned off by the absorbers.

Further, the bilinear-hysteretic absorber reduces the maximum response of the lower storey over a range of periods of the corresponding single-storeyed structures. Incidentally, this is the range of periods where, the maximum response of the one-storey system is high and amplitude reduction is desirable. The study also raises the problem of designing the absorber to withstand such large inputs without collapse. It may also be remarked that the inelastic vibration absorber concept, although found useful for the Taft S69°E ground motion, needs to be further explored with reference to different types of earthquake inputs.

9 Inelastic Torsional Response of a Single-Storey Framed Structure: Two Degree of Freedom System

9.1 INTRODUCTION

Buildings unsymmetric in plan are very common in practice. In most of the buildings there will be always distance of separation between center of mass and center of stiffness, or also called center of rigidity. It is called eccentricity. There will always be unsymmetry in either stiffness or mass. However, unsymmetry in mass will not be very apparent or visible and therefore is not identifiable. Therefore, it could be generally assumed to be symmetric because it is difficult to be controlled. On the other hand, unsymmetry in stiffness will be easily identifiable. Therefore, it is possible to adjust the lateral stiffness of various columns and shear walls in such a way that the eccentricity will be minimized.

IS 1893:2016 has very categorical recommendation against such eccentric buildings. IS 1893:2016 states that principal torsional mode should have a period, much shorter than the two other translational periods. Nextly, the ratio of the translational displacements of any two corners containing maximum and minimum values should not be more than 1.5.

Even about the mass participation factors, the IS 1893:2016 states that the mass participation factors for the principal torsional mode shall be pretty low.

All the above specifications are to restrict the eccentricity in plan. The torsion in plan increases the translational displacements of the columns in proportion to their distance from the mass center. The farthest column will suffer largest translational displacement due to torsion in plan. Due to the same reason, buildings that are very long and narrow in plan or whose aspect ratio is very large are not allowed.

In the author's research starting from the work for his PhD on inelastic torsional response, it was observed that ductility requirements are much larger, in columns farther away from the mass center. Therefore, it is really important to keep the eccentricity low. The author has also observed that it is possible to build a part of the building at one corner in plan such that the columns yield much earlier. The columns of such a corner building will be designed to supply large values of ductility. It will be much safer to build a soft and also a weak storey at one corner in plan than on top of the building. The weak corner storey can absorb the energy from the earthquake as much as the top expendable storey and still be safer than the top one with regards to its potential to cause damage to safety of people.

The author, while teaching the subject way back at IISc, had a personal experience of torsion in plan. He was travelling in a city bus a distance of about 15 km. The author was standing carrying a shoulder bag containing books weighing about 1 to 2 kg. As he was moving inside the bus towards the front, the driver suddenly applied the brakes. It took the author by surprise. Within a split second, the author was turned 180 degrees and was moved forward. He fell on the engine (inside the bus) backwards with his face upwards. It caused a small injury on the back of his head. Later the author analyzed the cause of the 180-degree turn and fall. It was nothing but the case of torsion in plan. As the author was carrying a relatively heavy bag on his side, there was an eccentricity of mass in plan. The sudden stopping of the bus caused a large eccentric inertia force in plan at the mass center which caused the rotation of the body in plan by 180 degrees. Of course, the shoulder twist of the body loosened his grip on the top bar of the bus which caused his fall.

The possibility of torsional response of a structure during earthquake ground motions has been recognized and certain provisions have been made in aseismic building design code to take the torsional effect into account. It must, however, be noted that these provisions do not consider the dynamic and non-linear characteristics of structural response. The presence of torsional deformation in building has been noticed during many earthquakes. The present state of aseismic design practice clearly indicates the need for further information on the torsional behavior of buildings during earthquakes, considering the dynamic and inelastic characteristics.

Torsional motions may be developed in buildings subjected to earthquakes for several reasons. If a building has an asymmetric floor plan causing a separation of the mass center from the center of stiffness, lateral forces acting at the mass center can lead to torsion. Alliteratively, even if the building is symmetric in plan, rotational components of ground motion can generate torsional response. Inelastic behavior of the components of a structure also can lead to torsional rotation of the structure. Even apparently symmetric structures can undergo torsional motions, because of accidental eccentricities may arise due to uncertainties in estimation of structural stiffness and due to non-linearities in response.

9.2 EARTHQUAKE RESPONSE OF ELASTIC STRUCTURE WITH COUPLED TRANSLATIONAL AND TORSIONAL MOTIONS

Ayre gave an analytical method using the normal mode concept to compute the earthquake response of shear buildings, both symmetric and asymmetric in plane. Gibson,

Moody and Ayre have demonstrated the application of response spectra to determine the earthquake response of tall buildings idealized as shear-flexible cantilever beams. They assumed that the model had three degrees of freedom, viz. horizontal translations, along the two orthogonal directions and a rotation about a vertical axis through the mass center.

Newmark presented the response spectrum for a symmetric building subjected to rotational components of ground motion. From the response spectrum, it is possible to pick out the magnitude of response contributed by torsional component, knowing the ratio of the translational to torsional frequency. Response spectrum technique and normal mode method have been applied to asymmetric structures by many others.

Bustamante and Rosenblueth showed that dynamic eccentricity, defined as the ratio of dynamic torque to dynamic shear causing the torque, can be more than static eccentricity, prescribed by codes. Study by Newmark on torsional response of symmetric structure due to rotational component of ground motion revealed that the accidental eccentricity of 5%, stipulated by code, is reasonable for buildings with periods greater than 0.6 sec, or shear-wall buildings with periods greater than 1.0 sec. However, for structures with shorter periods greater than 0.2 sec, the accidental eccentricity can be of the order of 10%. Housner and Outinen found that traditional static procedures to account for the effects of dynamic eccentricities on the earthquake response of structures were not in conformity with the dynamic analysis. Shiga from a study of earthquake response of one-storey and five-storey, elastic asymmetrical structures found that large eccentricities cause strong modal coupling.

Skinner, Skilton and Laws examined one-storey and five-storey structures subjected to the El Centro, May 1940, ground motion. They observed that close translational and torsional periods in a structure with even a small eccentricity and light damping would result in severe torsional effects.

Housner also observed that strong modal coupling results in a significant increase in the response. He expressed the strength of modal coupling as a ratio of the eccentricity to the difference in the corresponding translational and torsional frequencies. It was noted that strong modal coupling is possible even in nominally symmetric structures or structures with a small eccentricity provided that the translational and torsional frequencies are nearly equal. The two frequencies can nearly be the same in buildings with a uniform dispersion of columns over the floor area. Buildings with central cores or peripheral shear walls as primary resistive elements have well separated frequencies. Such buildings would be preferable, as they do not possess a strong coupling for small eccentricities. For the 22-storey, San Diego gas and electric company building, he found that earthquake response at the perimeter increased by as much as 85% relative to that expected for uncoupled modes. This was contrasted with the 30% increase at the periphery due to a 5% accidental eccentricity as prescribed by the code. This reflected the significant increase of the response due to modal coupling.

Several other studies indicated the same conclusion. It was also observed that close frequencies result in a beating type of oscillation.

Exterior frames undergo maximum displacements due to torsional effects. Blume pointed out that exterior frames should be provided with more resistance. Douglas and Trabert compared the responses obtained analytically with the observations made during a ground motion generated by a nuclear explosion of a 22-storey building. They observed that there was an increase in the response levels in the exterior frames at mid-heights. The increase was about 30 to 40% relative to that at the mass center. At lower storeys, an increase of about 100% over that of the frame at the mass center was observed.

The orientation of the building with respect to the direction of the ground motion causes significant changes in the response. Ayre determined experimentally the two orthogonal critical directions of a sinusoidal ground motion causing maximum and minimum responses, respectively, of a one-storey building model. Many others also determined the critical directions for multi-storey structures subjected to various inputs such as nuclear explosion and earthquake ground motion.

Certain unusual structures such as a tower on a building have characteristics very similar to those of torsionally unbalanced buildings. 1% is found that if the tower is light in weight, the dynamic behavior of such a structure is similar to a building with a small eccentricity. If the natural frequencies of the main structure and those of the tower are close, a strong modal coupling is found to exist.

Most of the investigations reported in the literature on the earthquake response of asymmetric structures are concerned with linear elastic analysis. It is, however, well recognized that strong ground motions can lead to responses of structure in the inelastic ranges. In this chapter, an attempt has been made to understand the inelastic torsional behavior of single-storey framed structures under ground motions. This chapter deals with a framed building, long, and rectangular in plan. Its response under a single horizontal component ground motion is studied. Chapter 7 examines the response of a framed structure square in plan under two horizontal components of ground motion.

9.3 STRUCTURAL MODEL

Fig. 9.1 shows a one-storey building structure. It consists of a rectangular slab, (sides axb) rigid in its own plane supported on a series of parallel portal frames of rise h. The shorter side 'a' of the rectangular slab is parallel to the span of the portals. The frames are identified by the numerals 1 through 5. The structure is assumed to have an eccentricity 'e'. The center of stiffness, S_c of the frames and the mass center, 'G' of the slab lie on the Y-axis with the separation distance of 'e'. Each frame is assumed to possess a bi-linearly hysteretic force-displacement relationship shown in Fig. 9.2. The structure is assumed to have two degrees of freedom, viz. a translational

Single-Storey Framed Structure: Two Degree of Freedom System

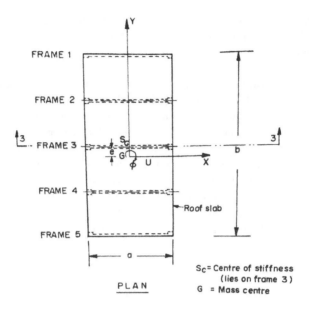

PLAN

S_c = Centre of stiffness
(lies on frame 3)
G = Mass centre

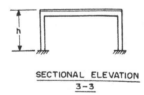

SECTIONAL ELEVATION
3-3

FIGURE 9.1: Structural model

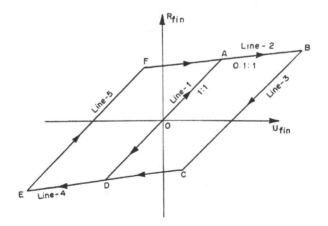

FIGURE 9.2: Force-displacement relationship

displacement, U of the mass center along the X-axis, and a rotational displacement Φ about a vertical axis through the mass center. Positive sense of the displacements is indicated in Fig. 9.1.

Each portal is assumed to deform only in the X-direction. The stiffness of each frame against a lateral load applied in the X-direction at the level of the roof slab has been determined. The effect of joint rotations of the portals has been taken into consideration. The total mass of the slab is assumed to be shared by the portal frames and is assumed to be lumped at the roof level. Linear elastic frequency of each frame has been computed from the values of a stiffness and mass. The yield displacement and the yield force of each frame at the roof level in the X-direction were determined from plastic analysis. Frames were assumed to yield, when the lateral spring force in the X-direction at the roof level attained its elastic limit. Fig. 9.3 shows the deformed structure. In all the discussions, the term 'building structure' means the entire structure

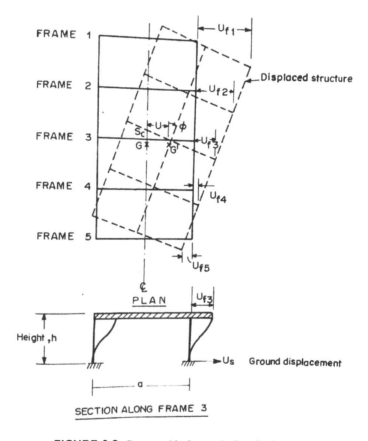

FIGURE 9.3: Structural before and after displacement

Table 9.1 Frequencies of structure

Frame period, T_{fi}	0.25 sec.		1.0 sec.	
	I Mode f_{t1} Hz.	II Mode f_{t1} Hz.	I Mode f_{t1} Hz.	II mode f_{t2} Hz.
e/b				
0	4.0	4.39	1.00	1.095
0.0125	3.98	4.40	0.995	1.100
0.025	3.96	4.45	0.989	1.110
0.05	3.84	4.55	0.960	1.140
0.10	3.58	4.90	0.895	1.220
0.20	3.10	5.65	0.775	1.410
0.25	2.92	6.05	0.730	1.515
0.30	2.71	6.46	0.678	1.615

Table 9.2 Frame properties

	0.25		1.0	
Frame period, T_{fi} sec.				
Frame yield acceleration q_{fiy}, g	0.14	0.07	0.14	0.07
Frame yield displacement U_{fiy}, cm	0.83	0.42	12.96	6.48

consisting of the portal frames and the roof slab. The term 'frame' refers to only portal frames. It has been assumed throughout that $b/a = 2.0$. Table 9.1 gives frequencies of the 1st and 2nd mode denoted by f_{t1} and f_{t2}, respectively. Frame properties are given in Table 9.2.

9.4 EQUATIONS OF MOTION

The two coupled equations of motion are,

$$M\ddot{U} + \Sigma_{i=1}^{N} R_{fi} - \Sigma_{i=1}^{N} \frac{m_i g}{h} U_{fi} + \Sigma_{i=1}^{N} C_{fi} \dot{U}_{fi} = -M\ddot{U}_S \qquad (9.1)$$

$$I\ddot{\Phi} + \Sigma_{i=1}^{N} R_{fi} r_{fi} - \Sigma_{i=1}^{N} \frac{m_i g}{h} U_{fi} r_{fi} - \Sigma_{i=1}^{N} C_{fi} \dot{U}_{fi} r_{fi} = 0 \qquad (9.2)$$

Here,

U = Translational displacement of the mass center along the direction of X-axis, relative to the ground.

Φ = Rotational displacement of the roof slab about a vertical axis through the mass center.

R_{fi} = Resistance of the ith frame - a function of the relative displacement, U_{fi} velocity, $\dot{U}_{fi}$ and time t.

$U_{fi} = U + r_{fi}\Phi$ - Displacement of the ith frame along the X-axis relative to the ground.

m_i = Mass supported by the ith frame.

g = Acceleration due to gravity.

h = Height of the portal frame.

C_{fi} = Damping constant of the ith frame.

U_S = Ground or support displacement along the X-axis.

M = Total mass of the complete structure.

I = Polar moment of inertia of the roof slab about a vertical axis through the mass center.

In all the expressions above, the subscript 'i' denotes the ith frame. The dot denotes differentiation with respect to time, t. It is to be noted that the 3rd term in Equations (9.1) and (9.2) constitutes the P-Δ effect.

The equations of motion may now be recast in non-dimensional form as follows:

$$Z'' + \Sigma_{i=1}^{N} R_{fin} S_{fir} Y_{fir} - \frac{Mg}{K_{f1}h} \Sigma_{i=1}^{N} \eta_{fir} Y_{fir} U_{fin} + 20 f_1 r \Sigma_{i=1}^{N} D_{fir} Y_{fir} U'_{fin} = -\frac{\ddot{U}_S}{Q_{f1y}}$$
(9.3)

$$\phi'' + R_{fin} S_{fir} Y_{fir} \frac{r_{fig}}{r_{gn}} - \frac{Mg}{K_{f1}h} \Sigma_{i=1}^{N} \eta_{fir} Y_{fir} U_{fin} \frac{r_{fig}}{r_{gn}} + 20 f_1 r \Sigma_{i=1}^{N} D_{fir} Y_{fir} U'_{fin} \frac{r_{fig}}{r_{gn}} = 0$$
(9.4)

$Z = \frac{U}{U_{f1y}}$ - Translational displacement of the mass center very along the X-axis normalized with respect to yield displacement of frame 1, U_{f1y}.

$R_{fin} = \frac{R_{fi}}{R_{fiy}}$ - Resistance of an ith frame normalized with respect to the yield resistance of the ith frame, R_{fiy}.

$S_{fir} = \frac{K_{fi}}{K_{f1}}$ - Ratio of stiffness of the ith frame to that of frame 1.

$Y_{fir} = \frac{U_{fiy}}{U_{f1y}}$ - Ratio of yield displacement of the ith frame to that of frame 1.

$\eta_{fir} = \frac{m_{fi}}{M}$ - Ratio of mass carried by the ith frame to total mass of the complete structure.

$U_{fin} = \frac{U_{fi}}{U_{fiy}}$ - Displacement of the ith frame along X-axis normalized with respect to yield displacement of the ith frame, U_{fiy}.

$C_{f1} = \frac{C_{f1}}{2\sqrt{K_{f1}M}}$ - Viscous damping ratio of frame 1.

$D_{fir} = \frac{C_{fi}}{C_{f1}}$ - Ratio of damping constant of the ith frame to that of frame 1.

$r_{fig} = \frac{r_{fi}}{r_g}$ - Distance of the ith frame from the mass center, normalized with respect to the polar radius gyration of the slab.

$r_{gn} = \frac{r_g}{U_{f1y}}$ - Polar radius of gyration of the slab about a vertical axis through the mass center normalized with respect to the yield displacement of frame 1.

Primes denote differentiation with respect to a non-dimensional time unit, τ, where
$\tau = \omega_{f1} t$

$\omega_{f1} = \frac{F_{f1}}{M}$

t = Time in seconds.

U_{f1y} = Yield displacement along X-direction of frame 1.

K_{f1} = Stiffness of frame 1.

C_{f1} = Damping constant of frame 1.

r_g = Polar radius of gyration of the slab about a vertical axis through the mass center.

$Q_{f1y} = \omega_{f1}^2 u_{f1y}$

q_{f1y} = acceleration or strength of frame 1.

q_{fiy} = acceleration or strength of frame i.

9.4.1 SOLUTION OF THE EQUATIONS OF MOTION

The equations of motion have been solved numerically by stepwise integration on an IBM 360/44 computer. A linear acceleration method has been employed for the integration. A step size of 1/80 of the fundamental period of the frame is used during the computation. (All the frames have been considered to have identical stiffness). All the frames are assumed to have identical damping constant equal to 1% of the critical.

A program in Fortran IV has been developed for the solution of the equations. The program is capable of tracing the hysteretic cycle of each frame individually. Two subroutines are included to facilitate the tracing of the different paths of the hysteretic cycle. The maximum ductility attained by each frame, the time at which it is attained, the maximum values of acceleration and velocity of each frame have been printed out. Besides the translational and rotational displacements, velocities and accelerations and their maximum values of the mass center have also been picked out. Detailed information on response-time histories, the total hysteretic energy absorbed up to the end of earthquake, total number of yield cycles have also been obtained.

9.4.2 PARAMETERS CONSIDERED IN THE STUDY

The main objective of the investigation is to find the effect of various structural and earthquake input parameters on the inelastic torsional response. The following parameters have been considered:

(i) Eccentricity, e
(ii) Linear elastic period of each frame, T_{fi}
(iii) Yield strength of each frame, q_{fiy}
(iv) Type of ground motion
(v) Non-Uniformity in the yield strength of the frames
(vi) P-Δ effect

ECCENTRICITY, E

The distance, 'e', between the mass center and the stiffness center has been varied by moving the mass center and keeping the stiffness center fixed at the geometric center of the structure. The shifting of the mass center is effected by a variation in the distribution of the mass. It has also been verified that the variation of the mass distribution is such as not to vary the value of the polar moment of inertia. The eccentricity has been expressed as a percentage of the length, 'b', of the building structure. Values of eccentricity ranging from 0 to 30% have been considered in the numerical studies.

LINEAR ELASTIC PERIOD OF EACH FRAME, T_{FI}

The linear elastic fundamental period of each frame has been varied from 0.25 sec through 3.0 sec. All the frames are assumed to have the same fundamental period. Response spectra for some values of eccentricity have been plotted. For only two values of the period, viz. 0.25 sec and 1.0 sec, the variation of eccentricity has been examined. The range of the period chosen is within the limits generally found for building structures. Although the longer periods are unrealistic for the single-storey structure, they have been included to study the effect of period.

YIELD ACCELERATION OF THE FRAMES, Q_{FIY}

Two values of yield acceleration, viz., 0.14 g and 0.07 g, have been considered for all the frames. All combinations of periods and yield accelerations are examined.

TYPE OF GROUND MOTION

Three different real earthquake accelerograms, viz., El Centro, May 1940, N-S component; Taft, July 1952, S69°E component; and El Centro, Borrego Mountain, April 1968, N-S component, have been used as input ground motions. The El Centro, May 1940, N-S component earthquake record has been widely used by many investigators because it is a typical strong ground motion record. The Taft, July 1952, S69°E component quake, which is about half as strong as the El Centro ground motion, has almost similar frequency characteristics as the El Centro record. The El Centro, Borrego Mountain, April 1968, N-S component accelerogram, which is still less intense compared to the Taft record, has a different frequency content.

The frequency content of an earthquake ground motion can be assessed by studying its acceleration response spectrum. From the acceleration response spectra with 2% damping for the El Centro, May 1940, N-S component earthquake, it is seen that there is a peak at a period equal to 0.44 sec. The spectra of Taft, July 1952, -S69°R component quake shows a peak at a period equal to 0.45 sec, while that of El Centro, Borrego Mountain, April 1968, N-S component ground motion shows a peak at 0.65 sec. Thus, the earthquakes of El Centro, May 1940 and Taft, July 1952 have input energy around higher frequencies, while the El Centro, Borrego Mountain, April 1968, has energy at lower frequencies.

Records of digitized values of these accelerograms up to the first 30 sec at intervals of 0.02 sec, published by the California Institute of Technology, have been used. Horizontal components of these earthquake accelerograms, only along X-axis (parallel to the shorter side of the building structure), have been used.

In the following pages of the thesis, the earthquakes at El Centro, May 1940, N-S component; Taft, July 1952, S69°E component; and El Centro, Borrego Mountain, April 1968, N-S component will be referred to simply as El Centro, Taft and Borrego Mountain earthquakes, respectively.

NON-UNIFORMITY IN THE STRENGTH OF THE FRAMES

The yield strength of the exterior frames 1 and 5 have been increased to 0.14 g, while that of the interior frames 2, 3 and 4 remained at 0.07 g. However, the elastic, linear fundamental period was identical in all the frames. The effect of doubling the strength of the exterior frames alone was examined for two values of the fundamental period, viz., $T_{fi} = 1.0$ sec and 0.25 sec.

P-Δ EFFECT

The majority of the results have been obtained by not considering the P-Δ effect. The P-Δ term has been included in four cases for the El Centro ground motion, viz.,

(i) $q_{fiy} = 0.14$, $T_{fi} = 0.25$ sec
(ii) $q_{fiy} = 0.14$, $T_{fi} = 1.0$ sec
(iii) $q_{fiy} = 0.07$, $T_{fi} = 0.25$ sec
(iv) $q_{fiy} = 0.07$, $T_{fi} = 1.0$ sec, to study its influence on the response.

9.4.3 DETAILS OF THE COMPUTER PROGRAM

Solution of the equations of motion yields the values of Z and Φ, the non-dimensional response values at the mass center. From these values, the response of each frame is computed by the formula $U_{fin} = (Z + r_{fin} \cdot \Phi)/Y_{fir}$ where U_{fin} is the displacement of each frame along the X-axis normalized with respect to its corresponding yield displacement, U_{fiy} and r_{fin} is the normalized distance of the ith frame from the mass center. In a similar way, the velocities and accelerations of each frame are computed.

The maximum excursion of U_{fin} during a cycle is defined as the ductility factor of the ith frame and is denoted by D_{fi}. The maximum value of D_{fi} attained over the duration of an earthquake is denoted by $\overline{D_{fi}}$.

9.5 DISCUSSION OF RESULTS

The results of the numerical studies have been presented in the form of graphs and tables. The response parameters used in the discussion are:

(i) The hysteretic energy absorbed by each frame.
(ii) The maximum frame ductility, $\overline{D_{fi}}$.

The influence of structure parameters like eccentricity, yield strength and fundamental period have been examined.

9.5.1 INFLUENCE OF ECCENTRICITY ENVELOPES OF MAXIMUM FRAME DUCTILITY

Figs. 9.4 to 9.6 refer to responses for the El Centro and Taft earthquakes and Fig. 9.6 refers to the Borrego Mountain earthquake. It is interesting to note that the responses

Single-Storey Framed Structure: Two Degree of Freedom System

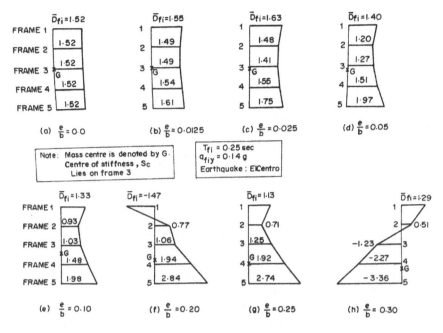

FIGURE 9.4: Maximum frame ductility, $\overline{D}_{fi}$

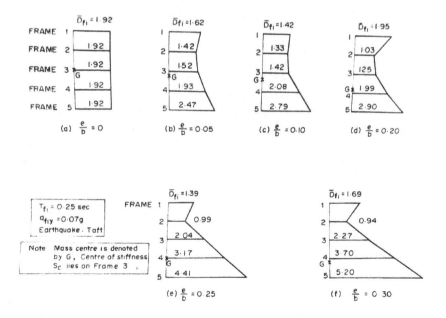

FIGURE 9.5: Maximum frame ductility, $\overline{D}_{fi}$

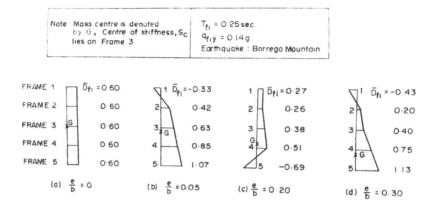

FIGURE 9.6: Maximum frame ductility, $\overline{D}_{fi}$

to El Centro and Taft quakes show significant inelastic behavior in most of the frames. The response to Borrego Mountain is, however, predominantly in the elastic range. Further, it may be noted that in Figs. 6.4 and 6.5, small eccentricities (like e = 0.05b) do not lead to significant departure from the no-torsion case (e = 0.0). It appears that, when the response is predominantly inelastic, the presence of a small eccentricity does not lead to a strong variation in the maximum displacements of the five frames. However, from Fig. 9.3, where the response is by and large elastic, for e = 0.05b, the variations of maximum ductility among frames shows a strong departure from the no-torsion case. The ratio of maximum ductilities of the two exterior frames is seen to increase with eccentricity when the response is inelastic. When the response is mostly elastic, this ratio of maximum ductilities does not seem to change significantly when the eccentricity is varied between 0.05b and 0.30b.

Fig. 9.7 shows the variation of maximum frame ductilities for all the five frames against eccentricity. The figure, again, confirms the wide variation of frame ductilities for large eccentricities, when the response is inelastic. In contrast, the ductility variation for the Borrego Mountain quake remains generally the same for small and large eccentricities. It may be noted that the response in this case is mostly elastic, since, practically all the ductilities are less than unity. It may be observed, in general, that the exterior frames can experience large ductilities, especially if the eccentricity is significant. This behavior needs to be taken into account while designing structures where eccentricity cannot be avoided.

9.5.2 INFLUENCE OF YIELD STRENGTH

The effect of a change in the yield strength of the frames is presented in Tables 9.3 to 9.5. Tables 9.3 and 9.4 refer to structures with a fundamental period of 0.25 sec and Table 9.5 refers to a structure with a period of 1.0 sec. Two values of yield strength of the frames are considered (q_{fiy} = 0.14 g and q_{fiy} = 0.07 g). Table 9.3 shows the

Single-Storey Framed Structure: Two Degree of Freedom System

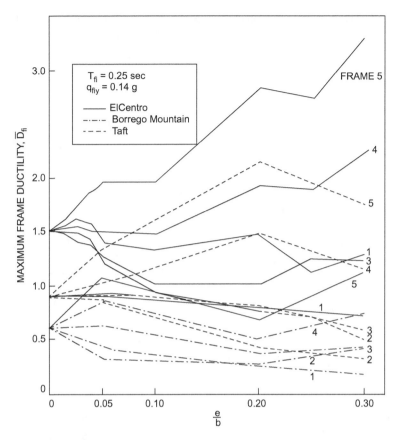

FIGURE 9.7: Variation of maximum frame ductility, $\overline{D}_{fi}$ with eccentricity

Table 9.3 Maximum ductility values, $\overline{D_{fi}}$ case with $q_{fiy} = 0.14g$ and $q_{fiy} = 0.07g$ are compared $T_{fi} = 0.25 sec$; El Centro earthquake

e/b	0		0.05		0.30	
q_{fiy}	0.14 g	0.07 g	0.14 g	0.07 g	0.14 g	0.07 g
Frame						
1	1.52	2.62	1.40	3.27	1.29	3.69
2	1.52	2.62	1.20	2.70	0.51	2.48
3	1.52	2.62	1.27	2.73	-1.23	4.36
4	1.52	2.62	1.51	3.58	-2.27	7.11
5	1.52	2.62	1.97	4.45	-3.36	10.55

Table 9.4 Maximum ductility values, $\overline{D}_{fi}$ case with $q_{fiy} = 0.14g$ and $q_{fiy} = 0.07g$ are compared $T_{fi} = 0.25sec$; Taft earthquake

Frame	e/b q_{fiy}	0.0 0.14 g	0.07 g	0.05 0.14 g	0.07 g	0.20 0.14 g	0.07 g	0.30 0.14 g	0.07 g
1		0.89	1.90	0.90	1.65	−0.79	1.95	−0.72	1.69
2		0.89	1.90	0.87	1.42	0.43	1.03	0.33	0.94
3		0.89	1.90	0.93	1.52	0.81	1.25	0.59	2.27
4		0.89	1.90	1.01	1.93	1.49	1.99	1.17	3.70
5		0.89	1.90	1.34	2.47	2.16	2.90	1.79	5.20

Table 9.5 Maximum ductility values, $\overline{D}_{fi}$ case with $q_{fiy} = 0.14g$ and $q_{fiy} = 0.07g$ are compared $T_{fi} = 1.0sec$; El Centro earthquake

Frame	e/b q_{fiy}	0.0 0.14 g	0.07 g	0.05 0.14 g	0.07 g	0.30 0.14 g	0.07 g
1		−1.25	1.60	1.10	1.55	−0.98	2.15
2		−1.25	1.60	0.79	1.46	−0.63	1.41
3		−1.25	1.60	−0.85	1.46	0.52	0.94
4		−1.25	1.60	−1.09	1.46	0.76	1.20
5		−1.25	1.60	−1.52	1.78	1.03	1.50

results for El Centro, May 1940, N-S component quake. It is interesting to note that when e/b = 0.0, the change in strength from 0.14 g to 0.07 g increases the ductility by 75%. However, as eccentricity increases, a reduction in strength strongly influences the ductility. In particular, for e = 0.35b, the exterior frame ductilities are increased by more than 200%. In Table 9.4, where the results for the Taft, July 1952 quake are presented, the yield strength does not seem to have as strong an influence as in the El Centro case. The general trend of the increase in ductility demand with a reduction in strength is, however, maintained.

In Table 9.5, which presents the data for the El Centro, 1940 Earthquake, the period being 1.0 sec, the results are quite different. The decrease in strength from 0.14 g to 0.07 g and leads only to marginal increases in ductility demand in most of the cases. Although, in most of the cases, frame 5 experiences the maximum ductility, it may be seen that in this case, for e = 0.30b, frame 1, rather than frame 5 suffers the maximum ductility, when the yield strength $q_{fiy} = 0.07$ g. The increase in period from 0.25 sec to 1.0 sec leads to smaller ductilities, for the El Centro, 1940 earthquake.

It may be mentioned, in general, a reduction in strength leads to increased ductility demand. The increase is, however, strongly influenced by the fundamental period and the type of earthquake.

9.5.3 INFLUENCE OF P-Δ EFFECT

The responses of the structure with and without P-Δ effect are compared in Tables 9.6 to 9.9. P-Δ effect seems to have negligible influence on the ductility requirements, for the structure parameters and earthquakes considered. It may be seen, from the tables, that structures with 1.0 sec periods are affected to a lesser extent by P-Δ effect than structures with 0.25 sec period. For the latter case, the change due to P-Δ effect is rarely more than 25%. It is interesting to note that P-Δ effect can lead to reduced ductilities in some cases.

9.5.4 INFLUENCE OF STRENGTHENING THE EXTERIOR FRAMES

It is already observed that the exterior frames experience large ductility demands. Tables 9.10 and 9.11 show the influence of non-uniformity in the yield strength of frames. Two structures, with frame periods of 0.25 sec and 1.0 sec have been chosen, while the yield strength of the exterior frames is doubled. It is seen that when e = 0, the value of the maximum ductility for the exterior frames 1 and 5 is half that of the interior frames 2, 3 and 4, as may be expected.

The results show a reduction in the ductility demand of all the frames by strengthening the exterior frames of the structure for which e/b is not zero. The reduction in the ductility of the exterior frames is significant for both the eccentricities considered.

Table 9.6 Maximum ductility values, $\overline{D_{fi}}$ case with and without P-Δ effect are compared $q_{fiy} = 0.14g$, $T_{fi} = 0.25sec$; El Centro earthquake

Frame	e/b 0.0 Without P-Δ	With P-Δ	0.05 Without P-Δ	With P-Δ	0.30 Without P-Δ	With P-Δ
1	1.52	1.58	1.40	1.42	1.29	1.54
2	1.52	1.58	1.20	1.31	0.51	0.52
3	1.52	1.58	1.27	1.47	-1.23	1.41
4	1.52	1.58	1.51	1.75	-2.27	2.55
5	1.52	1.58	1.97	2.18	-3.36	3.77

Table 9.7 Maximum ductility values, $\overline{D}_{fi}$ case with and without P-Δ effect are compared $q_{fiy} = 0.07g$, $T_{fi} = 0.25sec$; El Centro earthquake

Frame	e/b 0.0 Without P-Δ	With P-Δ	0.05 Without P-Δ	With P-Δ	0.30 Without P-Δ	With P-Δ
1	2.62	2.40	3.27	2.57	3.69	3.24
2	2.62	2.40	2.70	2.15	2.48	2.00
3	2.62	2.40	2.73	3.07	4.36	4.10
4	2.62	2.40	3.58	4.27	7.4	7.02
5	2.62	2.40	4.45	5.49	10.55	10.06

Table 9.8 Maximum ductility values, $\overline{D}_{fi}$ case with and without P-Δ effect are compared $q_{fiy} = 0.14g$, $T_{fi} = 1.0sec$; El Centro earthquake

Frame	e/b 0.0 Without P-Δ	With P-Δ	0.05 Without P-Δ	With P-Δ	0.30 Without P-Δ	With P-Δ
1	−1.25	−1.31	1.10	1.12	−0.98	−0.97
2	−1.25	−1.31	0.79	0.79	−0.63	−0.62
3	−1.25	−1.31	−0.85	−0.85	0.52	0.50
4	−1.25	−1.31	−1.09	−1.09	0.76	0.73
5	−1.25	−1.31	−1.52	−1.49	1.03	−1.01

Table 9.9 Maximum ductility values, $\overline{D}_{fi}$ case with and without P-Δ effect are compared $q_{fiy} = 0.07g$, $T_{fi} = 1.0sec$; El Centro earthquake

Frame	e/b 0.0 Without P-Δ	With P-Δ	0.05 Without P-Δ	With P-Δ	0.30 Without P-Δ	With P-Δ
1	1.60	1.61	1.55	1.54	2.15	2.13
2	1.60	1.61	1.46	1.45	1.41	1.39
3	1.60	1.61	1.46	1.45	0.94	0.96
4	1.60	1.61	1.46	1.54	1.20	1.16
5	1.60	1.61	1.78	1.81	1.50	1.45

Single-Storey Framed Structure: Two Degree of Freedom System

Table 9.10 Maximum ductility values, $\overline{D}_{fi}$ response values are compared with those of a structure where, $q_{fiy} = 0.07g$ for all the five frames $T_{fi} = 0.25sec$ $q_{fiy} = 0.14g$ for exterior frames 1 and 5 $q_{fiy} = 0.07g$ for interior frames 2, 3 and 4; El Centro earthquake

Frame	e/b = 0		0.05		0.30	
q_{fiy} =	0.14 g, exterior 0.07 g, interior	0.07 g, all frames	0.14 g, exterior 0.07 g, interior	0.07 g, all frames	0.14 g, exterior 0.07 g, interior	0.079, all frames
1	2.32	2.62	1.03	3.27	1.48	3.69
2	4.64	2.62	-2.05	2.70	-1.40	2.48
3	4.64	2.62	-2.44	2.73	-3.63	4.36
4	4.64	2.62	-2.92	3.58	-6.45	7.11
5	2.32	2.62	-1.73	4.45	-4.73	10.55

For example, in the case where, T_{fi}, = 0.25 sec. (Table 9.10) the reduction in the $\overline{D}_{fi}$ values over that, where all the frames have the same yield strength, is 55% for frame 5 and 60% for frame 1. The ductility demands made on the interior frames 2, 3 and 4 are marginally reduced when e = 0.05b or 0.30b. However, for e = 0, the interior frames seem to suffer greater ductilities owing to the strengthening of exterior frames. Similar trends are noticed for a structure with frame period of 1.0 sec (Table 9.11).

9.5.5 RESPONSE HISTORY CURVES

The histories of frame ductility response are shown in Figs. 9.8 to 9.11. Fig. 9.8 refers to response history of all the five frames of the structure with e = 0.0125 by for the El Centro earthquake. It appears from the diagram that all the frames execute vibrations which are more or less in phase. The dominant frequency content of vibration is 3.96 Hz, which compares with the fundamental frequency of 3.93 Hz. It is evident that the fundamental mode dominates the response. Modal patterns and frequency ratios, resulted from a free vibration analysis of the structure, are given in Fig. 9.12 for reference.

The response history of the five frames of the structure with e/b = 0.30, for the El Centro quake is shown in Fig. 9.9. Here, motions of frame 1 and frame 2 are more

Table 9.11 Maximum ductility values, $\overline{D_{fi}}$ response values are compared with those of a structure where, $q_{fiy} = 0.07g$ for all the five frames $T_{fi} = 1.0sec$ $q_{fiy} = 0.14g$ for exterior frames 1 and 5 $q_{fiy} = 0.07g$ for interior frames 2, 3 and 4; El Centro earthquake

Frame	e/b 0 q_{fiy} = 0.14 g, exterior q_{fiy} = 0.07 g, interior	= 0.07 g, all frames	0.05 = 0.14 g, exterior = 0.07 g, interior	= 0.07 g, all frames	0.30 = 0.14 g, exterior = 0.07 g, interior	= 0.07 g, all frames
1	0.96	1.60	-0.92	1.55	-0.98	2.15
2	1.92	1.60	-1.52	1.46	-1.26	1.41
3	1.92	1.60	-1.58	1.46	0.96	0.94
4	1.92	1.60	-1.85	1.46	1.38	1.20
5	0.96	1.60	-1.18	1.78	0.92	1.50

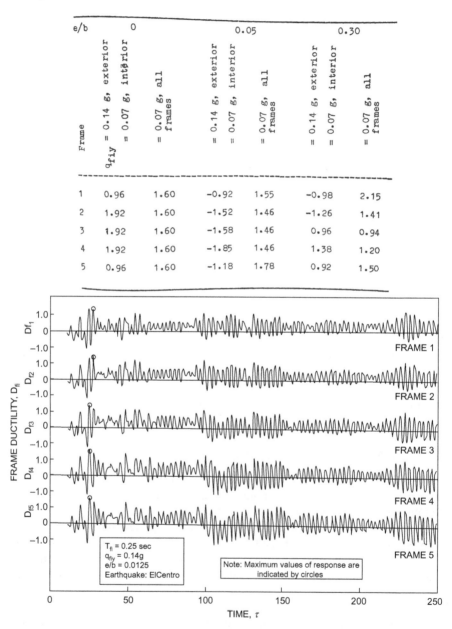

FIGURE 9.8: Response histories

Single-Storey Framed Structure: Two Degree of Freedom System

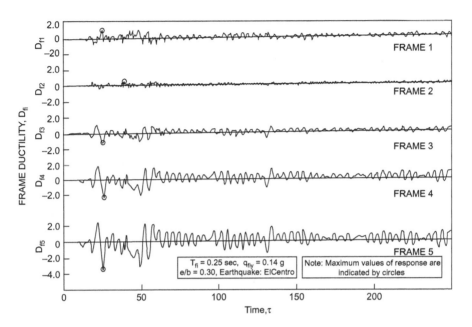

FIGURE 9.9: Response histories

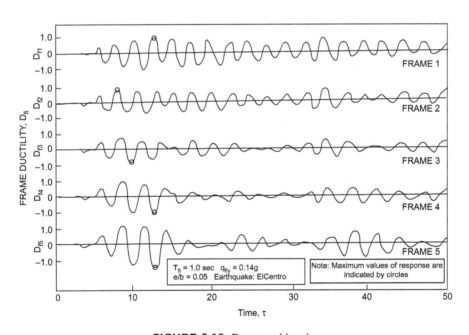

FIGURE 9.10: Response histories

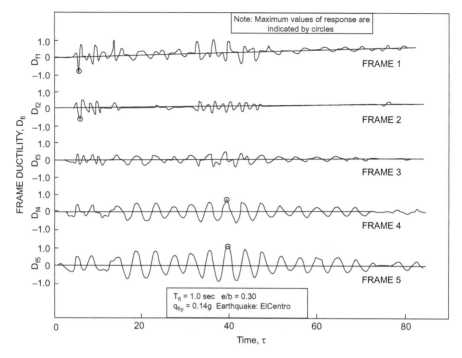

FIGURE 9.11: Response histories

or less out of phase. Frame 2 experiences relatively very small amplitude vibration and it is situated very close to the nodal point of I mode which can be observed from Fig. 9.12. The dominant frequency of motion is 2.64 Hz which is close to the fundamental frequency of 2.68 Hz. It may be noted here that, again the fundamental mode dominates the responses. For the structure with frame period equal to 1.0 sec and e = 0.05 b, the response history for the El Centro quake is shown in Fig. 9.10. The two linear frequencies for this structure are $f_{t1} = 0.96$ Hz and $f_{t2} = 1.15$ Hz. It is interesting to note that frame 5 executes a beating type of motion, which suggests significant participation of both the modes in the response. The dominant frequency of vibration of frame 5 is around 0.93 Hz, which is close to the fundamental frequency of 0.96 Hz. The frame 1, on the other hand, executes a vibration with a frequency of 1.13 Hz, which happens to be the frequency of the II mode, which is 1.15 Hz. A study of the mode shapes (Fig. 9.12) shows that the fundamental mode has hardly any amplitude at frame1. The first two modes of the structure are seen to participate significantly in the response. This behavior of the structure is, thus, totally in contrast to the behavior of the structure with frame period of 0.25 sec. The ductility response history of the five frames of the structure with frame period of 1.0 sec and e = 0.40b for the El Centro quake is shown in Fig. 9.11. The first two linear frequencies for this structure are 0.68 Hz and 1.63 Hz, respectively. The two frequencies are well separated because of the large eccentricity. It is seen that the dominant frequency of

Single-Storey Framed Structure: Two Degree of Freedom System

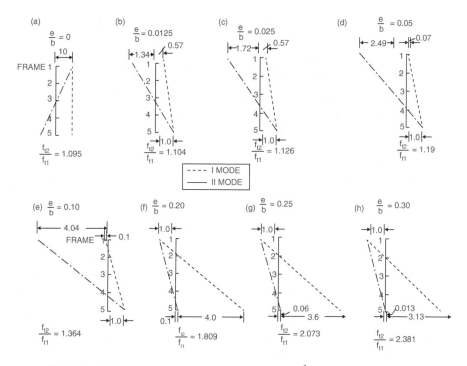

FIGURE 9.12: Modal patterns and frequency ratios, $\frac{f_{t2}}{f_{t1}}$ for eccentric structures

vibration of the frame 5 is 0.67 Hz. Hence the fundamental mode makes a significant contribution to the response of frame 5. Frame 1 undergoes a response which may be closely seen to be a superposition of motion in two frequencies. Obviously, both the modes influence the motion of frame 1. An interesting feature of the response is seen in the behavior of frame 2. The nodal point of the first mode coincides with the position of frame 2 and consequently, this frame remains quiescent when the II mode response is insignificant. The response of frame 2 consists of bursts of short duration vibration at a frequency of 1.63 Hz, interspersed with quiescent regimes. Thus, it may be seen that both the modes participate in the response of structure and the response of any particular frame is strongly influenced by the two modal patterns.

In summary, it may be said that the participation of the two modes of the structure is conditioned by (i) the fundamental period, (ii) the ratio of the first two frequencies, and (iii) frequency content of the particular earthquake input. The fundamental mode makes a significant contribution to the response for the structure with 4 Hz frame frequency. On the other hand, both the modes participate in the response when the frame frequency is 1 Hz. In the latter case, the two frequencies are in the significant frequency range as far as the El Centro quake is concerned and this explains the contribution by the two modes. It may be mentioned that when the 2nd mode participation is significant, the frame 1 can experience largest ductility relative to other frames (Fig. 9.12).

9.5.6 ENERGY DISSIPATION DUE TO HYSTERESIS

A study of the hysteretic energy dissipation leads to a finer understanding of the response of inelastic structures. With this object in view, the energy dissipated due to hysteresis has been computed. Figs. 9.13 and 9.14 show the variation of total hysteretic energy dissipated in each frame as a function of eccentricity. The values of energy (H_{fi}) have been non-dimensionalized with reference to the hysteretic energy dissipated (H_{fio}) in each frame for the zero-eccentricity case. The figures show that the frame 5 generally dissipates a very large proportion of hysteretic energy dissipated by the total structure. For instance, referring to Table 9.13, for the structure with frame period equal to 0.25 sec and Taft quake, frame 5 dissipates 68% of the total hysteretic energy dissipated by the structure as a whole, when e = 0.20b. It is, further, interesting to note that there is no simple correlation between the maximum ductility and the amount of hysteretic energy dissipation. For instance, as seen from Table 9.12 and Fig. 9.13, frame 5 dissipates maximum energy when e = 0.05b at a maximum ductility of 1.97, while the energy dissipated at e = 0.30 b is relatively less although the maximum ductility $\overline{D_{fi}}$ is 3.36 (Fig. 9.4). The reasons for this behavior are to be seen in Figs. 9.15 and 9.16, where the number of crossings of yield level are plotted against eccentricity. The close similarity between the two sets of figures suggests

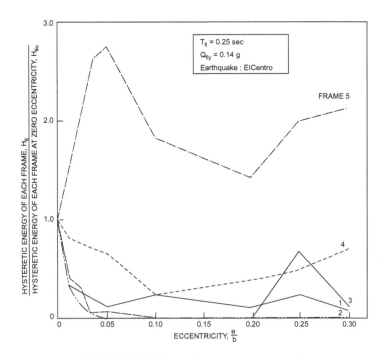

FIGURE 9.13: Hysteretic energy vs. eccentricity

Single-Storey Framed Structure: Two Degree of Freedom System

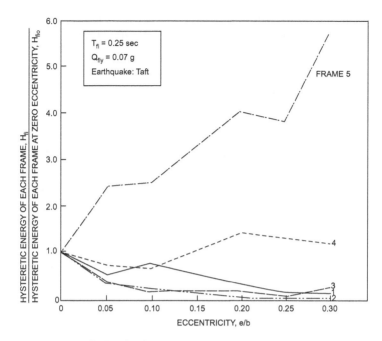

FIGURE 9.14: Hysteretic energy vs. eccentricity

Table 9.12 Distribution of hysteretic energy dissipation in frames. $T_{fi} = 0.25 sec$, $q_{fiy} = 0.14g$; El Centro earthquake

	Hysteretic energy dissipated in frame i				
	Total hysteretic energy dissipated by the structure				
e/b	Frame 1	Frame 2	Frame 3	Frame 4	Frame 5
0	0.2	0.2	0.2	0.2	0.2
0.05	0.03	0.0	0.020	0.18	0.77
0.10	0.10	0.0	0.0	0.10	0.80
0.20	0.04	0.0	0.0	0.18	0.78
0.25	0.06	0.0	0.2	0.14	0.60
0.30	0.02	0.0	0.03	0.23	0.72

that the total energy dissipated is influenced to a great extent by the number of yield level crossings. Figs. 9.17 to 9.19 show the time variation of the cumulative hysteretic energy, $H_{fc}(\tau)$ non-dimensionalized with respect to the elastic strain energy, E_L. Here, $E_L = \frac{1}{2} R_{fiy} \cdot U_{fiy}$ is the strain energy that can be stored elastically in a frame with yield

Table 9.13 Distribution of hysteretic energy dissipation in frames.
$T_{fi} = 0.25 sec$, $q_{fiy} = 0.07g$; Taft earthquake

Hysteretic energy dissipated in frame i

e/b	Total hysteretic energy dissipated by the structure				
	Frame 1	Frame 2	Frame 3	Frame 4	Frame 5
0	0.2	0.2	0.2	0.2	0.2
0.05	0.12	0.05	0.03	0.17	0.63
0.10	0.17	0.05	0.03	0.15	0.60
0.20	0.05	0.0	0.03	0.24	0.68
0.25	0.03	0.0	0.01	0.25	0.71
0.30	0.02	0.0	0.03	0.17	0.78

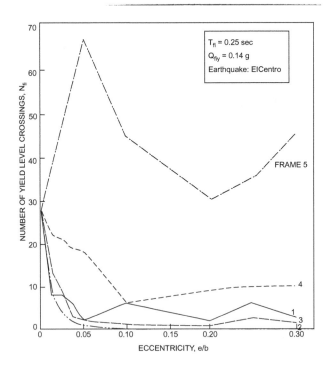

FIGURE 9.15: Number of yield level crossings vs. eccentricity

strength, q_{fiy} equal to 0.14. The cumulative energy function is seen to be monotonic and non-decreasing as may be expected. A major portion of energy is found to be dissipated during the first half of the earthquake duration. The frame 5 leads in the amount of energy dissipated throughout the duration of the quake.

Single-Storey Framed Structure: Two Degree of Freedom System

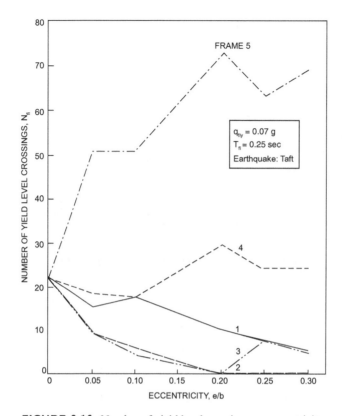

FIGURE 9.16: Number of yield level crossings vs. eccentricity

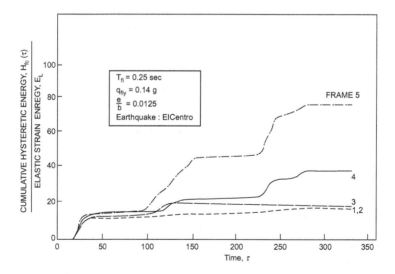

FIGURE 9.17: Cumulative hysteretic energy vs. time τ

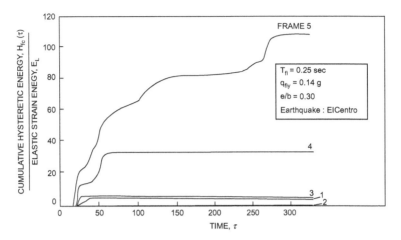

FIGURE 9.18: Cumulative hysteretic energy vs. time τ

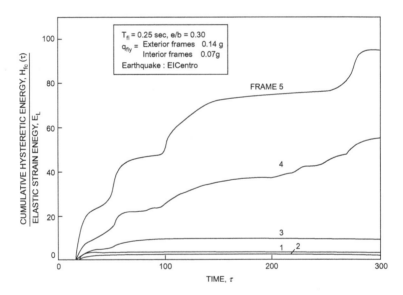

FIGURE 9.19: Cumulative hysteretic energy vs. time τ

Fig. 9.19 shows the time variation of $H_{fc}(\tau)$ for the structure, whose exterior frames have a yield strength twice those of the interior frames. Although the interior frame 4 experiences ductility greater than that of the exterior frame 5 by about 35%, the total hysteretic energy dissipated by the frame 5 is larger by about 75% compared with that of frame 4. It is interesting to see that, here also, the frame 5 is leading in the dissipation of energy through hysteresis. On comparing the Figs. 9.18 and 9.19, it can be observed that the magnitudes of hysteretic energy dissipated in all the frames are nearly the same in both the cases, viz. where the exterior frames have a yield strength of 0.14 g, while the interior frames have a yield strength of 0.07 g, and where all the frames have the same yield strength of 0.14 g.

9.5.7 MAXIMUM DUCTILITY RESPONSE SPECTRA

Spectra of maximum ductility response, $\overline{D_{fi}}$ of all the five frames of structure with $q_{fiy} = 0.14$ g, e = 0.0, 0.05 b and 0.30 b, respectively, are shown in Figs. 9.20, 9.21 and 9.22 for the El Centro ground motion. The ductility response in general seems to decrease with an increase in frame period. However, for e = 0 and 0.05 b, there is a sharp increase up to $T_{fi} = 0.5$ second and beyond which the $\overline{D_{fi}}$ decreases.

In general, frames 3, 4 and 5 follow the same pattern as the frame period is varied. When the frame period is between 0.5 sec and 1.2 sec, the exterior frame 1 undergoes relatively larger ductility response. When the period is 1.0 sec, the maximum ductility of frame 1 is next only to that of frame 5. Earlier, a study of response history curves

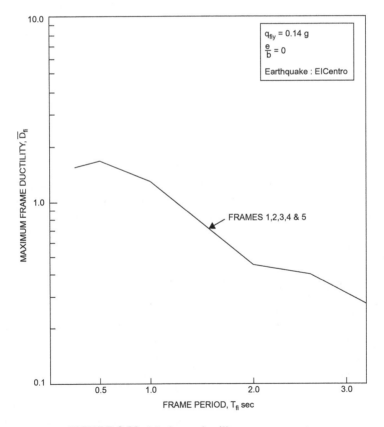

FIGURE 9.20: Maximum ductility response spectra

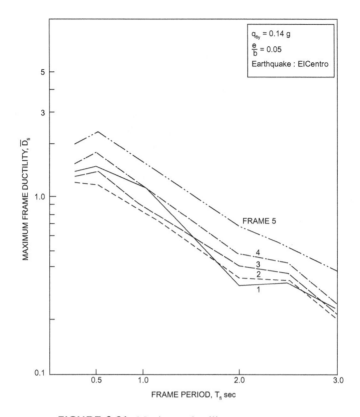

FIGURE 9.21: Maximum ductility response spectra

showed that the second mode participates significantly at a frame period of 1.0 sec. The spectra indicate that this feature may be expected for a range of periods around 1.0 sec. For values of frame period, T_{fi}, larger than 1.5 sec, all the frames respond in the elastic range. For some periods, the response of the frame 5 is even less than the response in the no-torsion case.

9.6 SUMMARY AND CONCLUSIONS

In this chapter, bilinear, hysteretic response of a one-storey building structure to several real earthquakes is discussed. The structure, consisting of a rectangular slab supported on a series of portal frames, is assumed to have two degrees of freedom. The several real earthquake accelerograms chosen for the study are, El Centro, May 1940, N-S component; Taft, July 1952, S69°E component; and El Centro, Borrego Mountain, April 1968, N-S component. It was intended to study the influence o various parameters like eccentricity, yield strength, linear elastic period, type of ground motion, non-uniformity in the strength of the frames on the response. P-Δ effect was also included during the course of this investigation. A nominal, 1% viscous

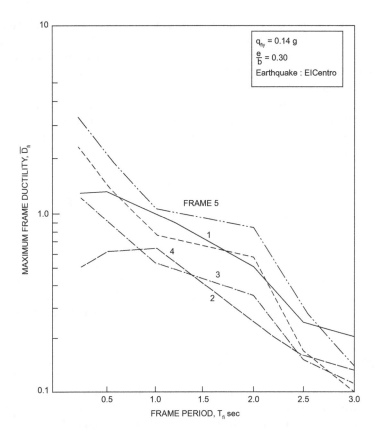

FIGURE 9.22: Maximum ductility response spectra

damping was assumed for the structure. The structure is assumed to deform in the direction of the input ground motion and, therefore, the frames were supposed to yield when the resistance force along the direction of displacement reached its yield value.

The structure responded in the inelastic range for El Centro and Taft ground motions. The response to Borrego Mountain quake was, however, mostly in the elastic range. For response situations which involve inelastic yielding, small eccentricities (up to 5%) do not lead to a significant change in the ductilities from the zero-torsion case. The effect of torsion becomes appreciable only for large eccentricities. However, for responses in the elastic range, even small eccentricities can lead to significant torsional behavior. Exterior frames, in general, suffered large ductility demands; the ratio of the ductilities of exterior frames was seen to increase with eccentricity when the response was inelastic. This ratio, however, remained more or less constant with eccentricity (from e = 0.05b onwards), when the response was elastic. This ratio however, remained more or less constant with eccentricity (from e = 0.05b onwards) when the response was elastic.

It is noted that when the yield strength of the frames decreased by half, the ductilities increased by nearly 75% for e/b = 0 and by 200% for e/b = 0.30., for structures with a frame period of 0.25 sec subjected to El Centro quake. However, structures with longer frame periods, like 1.0 sec, suffered only marginal increase in ductility due to the strength reduction. For Taft ground motion, the increase in the ductility due to a similar strength reduction was less, compared to that caused by El Centro motion. It may be noted that a reduction in strength, in general, leads to increased ductilities, which is, however, influenced by eccentricity, fundamental period and the type of ground motion.

The influence of P-Δ effect was also studied and it was seen that it had negligible influence on the ductility requirements for the type of structures, ground motion considered. Structures with shorter periods seem to have been affected more by P-Δ effect.

The effect of doubling the strength of only the exterior frames was seen to decrease the ductility demands of the exterior frames by nearly 60%, at eccentricities other than zero. However, a slight reduction in the interior frame ductilities was also observed at these eccentricities. When e/b = 0, the strengthening of the exterior frames caused an increase in the interior frame ductilities, although there was a marginal decrease in the ductility demands of the exterior frames.

It was also noted that the contribution made by the two modes of the structure was conditioned by the fundamental period, ratio of the two frequencies of the structure and the frequency content of the quake. In situations where the fundamental mode response was dominant, the exterior frame closer to the mass center experienced maximum ductility. When both the modes contributed significantly to the response, any of the two exterior frames could experience the maximum ductility.

Observations made on hysteretic energy dissipation revealed that no simple correlation can exist between the maximum ductility of a particular frame and the hysteretic energy dissipated by it. It was found that the total hysteretic energy dissipated by a frame is strongly influenced by the number of times it crosses the yield level. In the particular case of the structure with 0.25 sec. frame period, the exterior frame close to the mass center dissipated more than 60% of the total energy dissipated by the structure as a whole. Thus, it appears that the severity of the effect of eccentricity is revealed more by the hysteretic energy dissipation rather than the maximum ductility.

10 Inelastic Torsional Response of a Single-Storey Framed Structure: Three Degree of Freedom System

10.1 INTRODUCTION

Earthquake response of space structures is often computed by selecting a planar model of the structure and subjecting it to one of the components of earthquake ground motion. Interactions between various forces, which occur in a space structure are often ignored. Further, when the structure becomes inelastic during strong ground motion, the yielding pattern is altered due to the interactions. Besides, superposition of the responses, due to two components of ground motion acting independently, is not valid in the inelastic response studies. Thus, there is a need for a refined model for a structure, taking the interactions of various forces into account, which could lead to a finer understanding of structural response to earthquakes.

Nigam (1970) proposed a yielding model incorporating the effects of inelastic interaction between a set of six forces assumed for an element, viz. axial forces, shear forces, torsional moments and bending moments. He studied a series of elasto-plastic, single-storey structures supported on four columns subjected to two components of ground motion. The system had two degrees of freedom and torsion was not considered. Interactions between the bending moments in the two orthogonal directions were considered during the study. He assumed the yield surface to be a circle, and further it was assumed that the element goes into plastic state from elastic state without undergoing partly elastic and partly plastic state.

He compared the steady-state responses for three cases, viz., elastic, elasto-plastic without interaction and elasto-plastic with interaction. In the case of elasto-plastic without interaction, the inter-dependence of the response in X-direction on that in the Y-direction was neglected. He observed that the interaction had the effect of softening the system, reducing the amplitude and causing yielding at lower levels. It was also observed that the amplitudes of oscillation were not equal in two orthogonal directions when interaction was considered. Based on this, he indicated that interaction causes redistribution of energy in the two directions, so that the amplitude in one direction is increased and the amplitude in the perpendicular direction is decreased. It was noticed that interaction was important for short period structures only. He noted that hysteresis loop in one dimension differed from that in the biaxial case with interaction, since yield levels would be dependent upon the displacements in the two directions.

Kobori, Minai and Fujiwara (1973) presented a method for inelastic earthquake response analysis of multi-storey, multi-bay space frame structures with elasto-plastic joints. The inelastic responses to two components of El Centro ground motion were compared with those for one component of the quake. Nigam's yielding model was followed in their study. They observed that if the column members remained elastic or slightly plastic for the one component of the ground motion, the interaction between the two bending moments in the two-component case increases the ductilities. However, if the column behaved strongly plastically in the one-component case, the interaction effects do not always increase the ductilities in the two-component case because of the balancing of the ductilities in the two directions.

Pecknold (1974), investigated a series of elasto-plastic single-storey and three-storey structures subjected to several earthquakes. He compared the effect of simultaneous action of the two components of ground motion with that of only one component on the ductility requirements of the systems. The model assumed was a two degrees of freedom system and torsional motion was neglected. P-Δ effect was included in the study. He observed that for single-storey, elasto-plastic systems, proportioned according to the uniform building code, biaxial motion (with interaction) approximately doubles the ductility demands for stiff structures with period less than or equal to 0.3 sec. For longer fundamental periods, the increase in the ductility demand due to biaxial motion compared with the ductility for uniaxial motion is less.

In this chapter, a single-storey, bilinear, hysteretic structure, square in plan and supported on four columns, subjected to two horizontal ground motions is studied. The model is assumed to possess three degrees of freedom, viz., translational displacements along the two horizontal orthogonal directions and a rotation about the vertical axis. Interaction of the bending moments in the two perpendicular directions has been considered.

10.2 STRUCTURAL MODEL

Fig. 10.1 shows a one-storey building structure. It consists of a square Slab of side a rigid in its own plane, supported on four columns. The columns are identified by the numerals 1 through 4. The structure is assumed to have an eccentricity with respect to both X- and Y-axes. The eccentricity is caused by separation of the mass center from the center of stiffness. The distances of the mass center from the center of stiffness along the X- and Y-axes are denoted by e_x and e_y, respectively. Each column is circular in cross-section and is assumed to possess a force displacement relationship, which is bilinearly hysteretic in the uniaxial case (Fig. 10.2). The biaxial bending behavior of the column has been considered in arriving at the yield force of the column.

Single-Storey Framed Structure: Three Degree of Freedom System

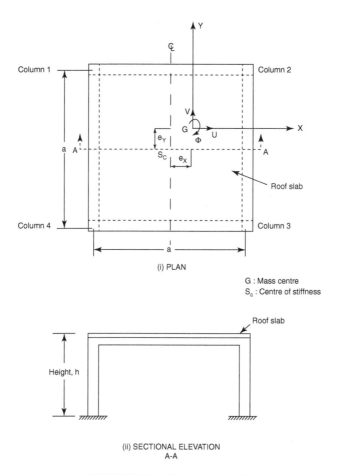

FIGURE 10.1: Structural model

The structure has three degrees of freedom, viz. translational displacements, U and V, of the mass center, along X- and Y-axes, respectively, and rotational displacement Φ, about a vertical axis through the mass center. The positive directions of the displacements are indicated in Fig. 10.1. The deformed structure is shown in Fig. 10.3.

10.2.1 YIELDING BEHAVIOR

It is assumed that the columns yield only at the ends. The column is said to have yielded when both the ends have yielded. The columns have been designed to carry biaxial moments. The yield displacements in the two orthogonal directions are equal

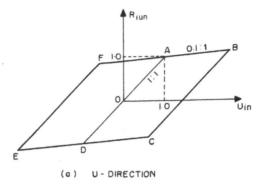

(a) U - DIRECTION

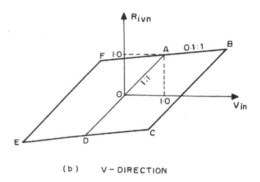

(b) V - DIRECTION

FIGURE 10.2: Force-displacement relationships for the uniaxial case

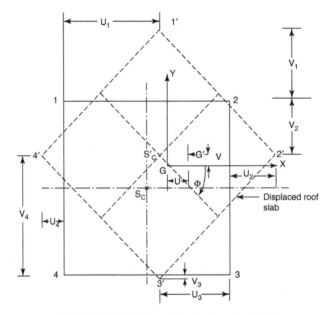

FIGURE 10.3: Plan of the displaced roof slab

Single-Storey Framed Structure: Three Degree of Freedom System

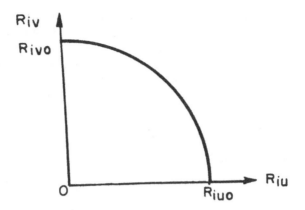

FIGURE 10.4: Yield surface

because the cross-section of the column is circular and thus symmetrical. Hence, the yield surface degenerates to a circle (Fig. 10.4) given by

$$\left(\frac{R_{iu}}{R_{iuo}}\right)^2 + \left(\frac{R_{iv}}{R_{ivo}}\right)^2 = 1$$

Where, R_{iu} = Resistance force of the column along the X-direction (corresponding to U-displacement).

R_{iuo} = Yield resistance force of the column along the X-direction (corresponding to the U-displacement).

R_{iv} = Resistance force of the column along the Y-direction (corresponding to V-displacement).

R_{ivo} = Yield resistance force of the column along the Y-direction (corresponding to the V-displacement).

Here, interaction of the axial load with the two biaxial moments is not considered. The hardening behavior in two-dimensional plasticity is generally considered in two or three ways. For isotropic hardening systems, the yield surface may be considered to expand uniformly to take the hardening into account. The Bauschinger effect is generally considered by assuming that the yield surface translates. In a more general situation both the effects may be included by considering expanding and translating yield surfaces. However, with the above considerations, the dynamical computations would become quite involved.

In the present study, the velocity reversals are assumed to take place independently in the two directions. This approximate and simple approach has been used in preference to a detailed consideration of changing yield surfaces. This artifice was resorted to on account of the extreme complexities of handling strain hardening in the biaxial bending analysis of a three degrees of freedom system. In other words, the biaxial interaction considered in an approximate way. It is believed that this approach will lead to a broad understanding of the interaction between torsion and two-dimensional motion, in spite of the severity of the approximation.

10.3 EQUATIONS OF MOTION

The three coupled equations of motion are,

$$M\ddot{U} + \sum_{i=1}^{N_C} R_{iu} + C_i \sum_{i=1}^{N_C} \dot{U}_i = -M\ddot{U}_S \qquad (10.1)$$

$$M\ddot{V} + \sum_{i=1}^{N_C} R_{iv} + C_i \sum_{i=1}^{N_C} \dot{V}_i = -M\ddot{V}_S \qquad (10.2)$$

$$I\ddot{\Phi} + \sum_{i=1}^{N_C} R_{iu} r_{iv} - \sum_{i=1}^{N_C} R_{iv} r_{iu} + C_i \sum_{i=1}^{N_C} \dot{U}_i r_{iv} - C_i \sum_{i=1}^{N_C} \dot{V}_i r_{iu} = 0 \qquad (10.3)$$

Here,

U, V = translational displacements of the mass center along the directions of X-axis and Y-axis, respectively, relative to the ground.

Φ = rotational displacement of the slab about a vertical axis, through the mass center.

R_{iu}, R_{iv} = resistance of the ith column along the X- and Y-axes, respectively - functions of relative displacements, relative velocities and time.

$U_i = U + r_{iv}\Phi$ - displacement of the ith column along the X-axis.

$V_i = V + r_{iu}\Phi$ - displacement of the ith column along the Y-axis.

r_{iu}, r_{iv} = distances of the ith column from the mass center along X- and Y-axes, respectively.

C_i = damping constant of the ith column.

J_S, V_S = ground or support displacements along the X- and Y-axes, respectively.

M = mass of the complete structures assumed to be lumped at the floor level.

I = polar moment of inertia of the roof slab about a vertical axis through the mass center.

N_C = number of columns.

Single-Storey Framed Structure: Three Degree of Freedom System

In all the expressions, the subscript 'i' refers to the ith column. The dot denotes differentiation with respect to time 't'.

The equations of motion may now be recast in a non-dimensional form as follows:

$$U_n'' + \Sigma_{i=1}^{NC} R_{iun} + 2S_{ic}\Sigma_{i=1}^{NC} U_{in}' = \frac{-\ddot{U}_S}{Q_{iuo}} \quad (10.4)$$

$$V_n'' + \Sigma_{i=1}^{NC} R_{ivn} + 2S_{ic}\Sigma_{i=1}^{NC} V_{in}' = \frac{-\ddot{V}_S}{Q_{ivo}} \quad (10.5)$$

$$\Phi'' + \Sigma_{i=1}^{NC} R_{iun}\frac{r_{igv}}{r_{gnu}} - \Sigma_{i=1}^{NC} R_{ivn}\frac{r_{igu}}{r_{gnv}} + 2S_{ic}\Sigma_{i=1}^{NC} U_{in}'\frac{r_{igv}}{r_{gnu}} - 2S_{ic}\Sigma_{i=1}^{NC} V_{in}'\frac{r_{igu}}{r_{gnv}} = 0$$

Where,

$U_n = \frac{U}{U_{io}}$ - non-dimensional displacement of the mass center along the X-axis.

$V_n = \frac{V}{V_{io}}$ - non-dimensional displacement of the mass center along the Y-axis.

$R_{iun} = \frac{R_{iu}}{R_{iuo}}$ - normalized resistance of the ith column along the X-axis.

$R_{ivn} = \frac{R_{iv}}{R_{ivo}}$ - normalized resistance of the ith column along the Y-axis.

$U_{in} = \frac{U_i}{U_{io}}$ - normalized displacement of the ith column along the X-axis.

$V_{in} = \frac{V_i}{V_{io}}$ - normalized displacement of the ith column along the Y-axis.

$S_{ic} = \frac{C_i}{2\sqrt{K_{iu}M}}$ - non-dimensional damping ratio

(It is assumed that $S_{ic} = 0.01$, throughout the study.)

K_{iu}, K_{iv} = elastic stiffnesses of the ith column along the X- and Y-axes, respectively.

The columns are symmetrical in cross-section and therefore $K_{iu} = K_{iv}$.

$r_{gnu} = \frac{r_g}{U_{io}}$ - polar radius of gyration of the slab about a vertical axis through the mass center, normalized with respect to yield displacement of the ith column along the X-axis.

$r_{gnv} = \frac{r_g}{V_{io}}$ - polar radius of gyration of the slab about a vertical axis through the mass center, normalized with respect to yield displacement of the ith column along the Y-axis.

$r_{igu} = \frac{r_{iu}}{r_g}$ - normalized distance of the ith column from the mass center, measured along the X-axis.

$r_{igv} = \frac{r_{iv}}{r_g}$ - normalized distance of the ith column from the mass center, measured along the Y-axis.

$Q_{iuo} = \frac{R_{iuo}}{M}$
$Q_{ivo} = \frac{R_{ivo}}{M}$
q_{iuo}, q_{ivo} = yield accelerations or strengths of the ith column along X- and Y-axes, respectively.

U_{io}, V_{io} = yield accelerations of the ith column along X- and Y-axes, respectively, corresponding to uniaxial loading (It is assumed throughout $U_{io} = V_{io}$)

R_{iuo}, R_{ivo} = yield resistance of the ith column along X- and Y-axes, respectively, corresponding to uniaxial loading (It is assumed throughout $R_{iuo} = R_{ivo}$)

r_g = polar radius of gyration of the slab about a vertical axis through the mass center.

$u = \sqrt{\frac{K_{iu}}{M}}$
$v = \sqrt{\frac{K_{iv}}{M}}$
t = time in seconds.

$\tau = \omega_u t = \omega_v t$
Prime denotes differentiation with respect to non-dimensionalized time unit τ.

10.4 SOLUTIONS OF THE EQUATIONS OF MOTION

The equations of motion in the incremental form are numerically integrated by the linear acceleration method on the IBM 360/44 computer. A step size of 1/80 of the fundamental period of the ith column, T_i has been used during the computation.

A program in Fortran IV has been developed for the solution of the equations. The program is capable of tracing the hysteretic cycle of each column, in the two orthogonal directions. In addition to the two subroutines already mentioned in Chapter 9, one more subroutine to calculate the displacements U_i and V_i which satisfy the yield condition is included. Combination of the column displacements in the two orthogonal directions, which satisfies the yield condition is determined precisely. The transition from elastic to inelastic regime occurs during a finite time increment. The time at which the resistant forces cross the yield surface is noted and by using a finer time increment along with iteration, the yield displacement is exactly determined. Maximum ductility attained by each column, the time at which it is attained, the maximum values of acceleration and velocity of each column have been printed out. Besides, the translational and rotational displacements, velocities and accelerations of the mass center and their maximum values have also been printed out. Further, response histories have also been observed.

The maximum values of U_n and V_n attained by the ith column during a cycle of hysteresis are termed as ductilities, D_{iu} and D_{iv}, respectively, and the maximum values of these over the duration of the earthquake are called maximum ductilities, $\overline{D_{iu}}$ and $\overline{D_{iv}}$, respectively. Further, the largest of the maximum values De and Ds, of all the columns is denoted by $\overline{D_{max}}$ and the smallest of the maximum values $\overline{D_{iu}}$ and $\overline{D_{iv}}$ of the columns is denoted by $\overline{D_{min}}$.

10.4.1 PARAMETERS CONSIDERED IN THE STUDY

The influence of the following parameters on the inelastic torsional response has been investigated.

(i) Eccentricities, e_x and e_y, (It is assumed, throughout, that $e_x = e_y = e$). Eccentricity has been expressed as a fraction, e/a of the length (also width) of the roof slab.

(ii) Linear elastic period of each column, T_{iu}, T_{iv} ($T_{iu} = T_{iv}$ T_i, because of the symmetry of the cross-section of the column).

(iii) Yield strength of each column, q_{iuo}, q_{ivo} ($q_{iuo} = q_{ivo} = q_i$, because of the symmetry of the cross-section of the columns).

Eccentricity e/a has been varied from 0 to 0.30. Systems with T_i = 0.25 sec, 0.35 sec, and 1.0 sec have been studied. Two values of yield strength, q_i viz., 0.088 g and 0.044 g, where g is the acceleration due to gravity have been considered in the study. Tables 10.1 and 10.2 and Fig. 10.13 give the structure frequencies and system properties. Results have been obtained for the following two cases:

(i) Both the X and Y components of the ground motion are assumed to act on the structure simultaneously.

(ii) Only one of the two orthogonal horizontal components, viz. X-component, of the ground motion is taken to act on the structure.

For simplicity, the first case will be termed as a two-component case and the second one will be called a one-component case. The two orthogonal horizontal components of the El Centro, May 1940, ground motion have been considered for the study. Digitized values of the acceleration records for the first 15 sec of each of the orthogonal horizontal components of the quake, published by the California Institute of Technology have been employed. The NS component of the quake has been taken as the X-component of the ground motion, and the S90W component of the quake has been taken as the Y-component of the ground motion.

Table 10.1 Structure frequencies (Hz)

Period T_i sec.		0.25			0.35			1.0		
Eccentricity e/a	I Mode	II	III	I	II	III	I	II	III
0	4.0	4.0	6.95	2.86	2.86	5.0	1.0	1.0	1.74
0.0125	4.0	4.0	6.95	2.86	2.86	5.0	1.0	1.0	1.74
0.05	3.99	4.0	7.05	2.84	2.86	5.0	1.0	1.0	1.75
0.10	3.89	4.0	7.15	2.78	2.86	5.10	0.97	1.0	1.78
0.20	3.46	4.0	8.0	2.50	2.86	5.72	0.87	1.0	2.00
0.30	3.30	4.0	8.4	2.38	2.86	5.99	0.83	1.0	2.10

Table 10.2 System properties

Period, T_i sec.	Yield Strength q_i, g.	Yield displacement, $U_{io}(V_{io})$, cm
0.25	0.088	0.546
	0.044	0.273
0.35	0.088	1.050
	0.044	0.525
1.0	0.088	8.741
	0.044	4.370

10.5 DISCUSSION OF RESULTS

The results of the numerical studies are presented in the form of graphs and tables. The major response parameters studied are the maximum column ductilities $\bar{D}_{iu}$ and $\bar{D}_{iv}$. The influence of structure parameters like eccentricity, yield strength and fundamental period on the maximum column ductilities have been examined.

10.5.1 INFLUENCE OF ECCENTRICITY

Tables 10.3 through 10.7 give the maximum column ductilities $\bar{D}_{iu}$ and $\bar{D}_{iv}$. Ductility values of the structure when acted upon by both the X- and Y-components of the ground motion simultaneously (two-component case) and X-component of the ground motion (one-component case) are compared. It is interesting to see that, in both the cases, $\bar{D}_{1u}$ equals $\bar{D}_{2u}$, $\bar{D}_{3u}$ equals $\bar{D}_{4u}$, $\bar{D}_{1v}$ equals $\bar{D}_{4v}$ and $\bar{D}_{2v}$ equals $\bar{D}_{3v}$ at all

Single-Storey Framed Structure: Three Degree of Freedom System

Table 10.3 Maximum ductilities, $\overline{D}_{iu}, \overline{D}_{iv} T_i = 1.0 sec; q_i = 0.044g$ ground motion; El Centro; (i) two-component case (ii) one-component case

e/a	0		0.0125		0.05		0.10		0.25		0.30	
Column	(i)	(ii)	(i)	(ii)	(i)	(ii)	(i)	(ii)	(i)	(ii)	(i)	(ii)
1 $\overline{D}_{1u}$	-1.95	-2.09	-1.95	-2.14	-2.12	-2.28	-1.50	2.37	-1.68	2.14	-1.71	2.09
$\overline{D}_{1v}$	-3.09	0.00	-3.09	-0.39	-3.36	-1.23	-2.83	-1.72	-2.21	2.06	2.48	-2.33
2 $\overline{D}_{2u}$	-1.95	-2.09	-1.95	-2.14	-2.12	-2.28	-1.50	2.37	-1.68	2.14	-1.71	2.09
$\overline{D}_{2v}$	4.94	0.00	5.06	0.43	5.36	0.99	5.75	0.97	6.47	-1.09	6.64	-1.68
3 $\overline{D}_{3u}$	-4.00	-2.09	-3.95	-1.91	-3.75	-2.12	4.06	1.81	3.57	1.25	3.39	-1.21
$\overline{D}_{3v}$	4.94	0.00	5.06	0.43	5.36	0.99	5.75	0.97	6.47	-1.09	6.64	-1.68
4 $\overline{D}_{4u}$	-4.00	-2.09	-3.95	-1.91	-3.75	-2.12	4.06	1.81	3.57	1.25	3.39	-1.21
$\overline{D}_{4v}$	-3.09	0.00	-3.09	-0.39	-3.36	-1.23	-2.83	-1.72	-2.21	2.06	2.48	-2.33

Table 10.4 Maximum ductilities, $\overline{D}_{iu}, \overline{D}_{iv} T_i = 1.0 sec; q_i = 0.088g$ ground motion; El Centro; (i) two-component case (ii) one-component case

e/a	0		0.0125		0.05		0.10		0.25		0.30	
Column	(i)	(ii)	(i)	(ii)	(i)	(ii)	(i)	(ii)	(i)	(ii)	(i)	(ii)
1 $\overline{D}_{1u}$	-1.25	1.40	1.47	-1.49	-1.78	1.68	1.29	1.67	-1.24	1.27	-1.57	-2.55
$\overline{D}_{1v}$	-1.29	0.00	-1.24	-0.22	-1.21	-0.75	-1.57	-0.78	-1.37	-1.07	-1.23	-1.13
2 $\overline{D}_{2u}$	-1.25	1.40	1.47	-1.49	-1.78	1.68	1.29	1.67	-1.24	1.27	-1.57	-2.55
$\overline{D}_{2v}$	-2.22	0.00	-3.04	-0.24	2.27	0.80	-2.76	-0.68	-2.58	-0.93	2.33	2.19
3 $\overline{D}_{3u}$	-1.49	1.40	-1.69	1.32	1.22	1.09	1.29	0.84	1.34	-1.01	1.36	-1.02
$\overline{D}_{3v}$	-2.22	0.00	-3.04	0.24	2.27	0.80	-2.76	-0.68	-2.58	-0.93	2.33	2.19
4 $\overline{D}_{4u}$	-1.49	1.40	-1.69	1.32	1.22	1.09	1.28	0.84	1.34	-1.01	1.36	-1.02
$\overline{D}_{4v}$	-1.29	0.00	-1.24	-0.22	-1.21	-0.75	-1.57	-0.78	-1.37	-1.07	-1.23	-1.13

Table 10.5 Maximum ductilities, $\overline{D}_{iu}, \overline{D}_{iv} T_i = 0.35 sec; q_i = 0.044g$ ground motion; El Centro; (i) two-component case (ii) one-component case

e/a	0		0.0125		0.05		0.10		0.25		0.30	
Column	(i)	(ii)	(i)	(ii)	(i)	(ii)	(i)	(ii)	(i)	(ii)	(i)	(ii)
1 $\overline{D}_{1u}$	-10.42	-7.56	-10.44	-7.92	-10.11	-8.60	-10.65	-8.73	-10.34	8.00	-10.31	7.96
$\overline{D}_{1v}$	-14.00	0.00	-14.21	1.51	-15.17	-3.10	-16.99	-3.58	-19.98	-4.17	-20.56	-5.75
2 $\overline{D}_{2u}$	-10.42	-7.56	-10.44	-7.92	-10.11	-8.60	-10.65	-8.73	-10.34	8.00	-10.31	7.96
$\overline{D}_{2v}$	-25.77	0.00	-26.28	-1.99	-28.57	-2.81	-29.33	-2.24	-36.79	2.99	-38.71	3.16
3 $\overline{D}_{3u}$	-10.67	-7.56	-11.26	-7.18	-13.41	-5.92	-14.68	-5.28	-12.46	-5.34	-11.82	-3.28
$\overline{D}_{3v}$	-25.77	0.00	-26.28	-1.99	-28.57	-2.81	-29.33	-2.24	-36.79	2.99	-38.71	3.16
4 $\overline{D}_{4u}$	-10.65	-7.56	-11.26	-7.18	-13.41	-5.92	-14.68	-5.28	-12.46	-5.34	-11.82	-3.28
$\overline{D}_{4v}$	-14.00	0.00	-14.21	1.51	-15.17	-3.10	-16.99	-3.58	-19.98	-4.17	-20.56	-5.75

Table 10.6 Maximum ductilities, $\bar{D}_{iu}, \bar{D}_{iv} T_i = 0.35 sec; q_i = 0.088g$ ground motion; El Centro; (i) two-component case (ii) one-component case

e/a		0		0.05		0.10		0.25	
Column		(i)	(ii)	(i)	(ii)	(i)	(ii)	(i)	(ii)
1	$\bar{D}_{1u}$	-5.29	2.41	-4.79	2.89	4.36	3.10	-4.10	-2.69
	$\bar{D}_{1v}$	-4.69	0.00	-3.27	0.81	-2.82	1.50	-3.86	1.80
2	$\bar{D}_{2u}$	-5.29	2.41	-4.79	2.89	4.36	3.10	-4.10	-2.69
	$\bar{D}_{2v}$	-7.41	0.00	-10.08	-0.92	-9.89	-1.63	-8.40	-1.98
3	$\bar{D}_{3u}$	3.67	2.41	5.89	1.90	5.79	1.42	4.77	-1.75
	$\bar{D}_{3v}$	-7.41	0.00	-10.08	-0.92	-9.89	-1.63	-8.40	-1.98
4	$\bar{D}_{4u}$	3.67	2.41	5.89	1.90	5.79	1.42	4.77	-1.75
	$\bar{D}_{4v}$	-4.69	0.00	-3.27	0.81	-2.82	1.50	-3.86	1.80

Table 10.7 Maximum ductilities, $\bar{D}_{iu}, \bar{D}_{iv} T_i = 0.25 sec; q_i = 0.088g$ ground motion; El Centro; (i) two-component case (ii) one-component case

e/a		0		0.0125		0.05		0.10		0.25		0.30	
Column		(i)	(ii)	(i)	(ii)	(i)	(ii)	(i)	(ii)	(i)	(ii)	(i)	(ii)
1	$\bar{D}_{1u}$	-11.09	2.08	-10.86	2.16	-10.44	2.53	-9.81	2.69	-7.28	-3.41	-6.71	-3.92
	$\bar{D}_{1v}$	-6.61	0.00	-6.49	-0.28	-5.65	-0.92	-5.00	1.79	-6.14	-1.26	-6.43	-1.60
2	$\bar{D}_{2u}$	-11.09	2.08	-10.86	2.16	-10.44	2.53	-9.81	2.69	-7.28	-3.41	-6.71	-3.92
	$\bar{D}_{2v}$	-12.21	0.00	-12.23	0.29	-13.83	-1.19	-14.49	-1.56	-14.03	2.78	-14.27	2.22
3	$\bar{D}_{3u}$	8.92	2.08	8.70	2.04	8.38	1.89	7.73	1.59	-5.95	-2.85	-5.63	-3.12
	$\bar{D}_{3v}$	-12.21	0.00	-12.23	0.29	-13.83	-1.19	-14.49	-1.56	-14.02	2.78	-14.27	2.22
4	$\bar{D}_{4u}$	8.92	2.08	8.70	2.04	8.38	1.89	7.73	1.59	-5.95	-2.85	-5.63	-3.12
	$\bar{D}_{4v}$	-6.61	0.00	-6.49	-0.28	-5.65	-0.92	-5.00	1.79	-6.14	-1.26	-6.43	-1.60

eccentricities. This behavior is a result of the assumption that the slab is rigid in its own plane.

When the eccentricity is small, the maximum ductilities are not significantly different from those of the zero eccentricity case. When the eccentricity, e/a is of the order of 0.30, torsion does affect the maximum ductilities. While it reduces the maximum ductility for some columns, the largest value of maximum ductility, experienced by column 3 or 2 in the Y-direction ($\bar{D}_{2v} = \bar{D}_{3v}$) is generally increased due to eccentric torsion. The magnitude of the increase varies from 54% to 50% depending on the period and yield strength of the structure.

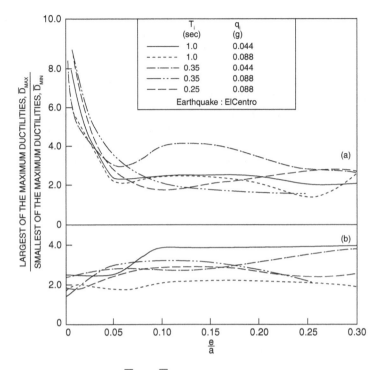

FIGURE 10.5: Variation of the $\overline{D}_{MAX}/\overline{D}_{MIN}$ with eccentricity: (a) for X-component of ground motion on the structure; (b) for both X- and Y-components of the ground motion on the structure

The ratio of the maximum of the maximum ductilities ($\overline{D}_{max}$) to the minimum of the maximum ductilities ($\overline{D}_{min}$) is seen to increase with increasing eccentricities, for the two-component case; this indicates that the disparity between the ductilities becomes larger for larger values of e/a. The ratio becomes smaller for larger eccentricities (Fig. 10.5) in the one-component case. At all eccentricities and in all the cases studied, the maximum ductility experienced by columns 2, 3 ($\overline{D}_{2v} = \overline{D}_{3v}$) in the Y-direction is the largest of all the other column ductilities. It is instructive to note that the X-coordinate of the columns 2, 3 with respect to the mass center is smaller than the X-coordinate of the columns 1, 4 and thus nearer to mass center. A Similar trend was noticed in the structure with two degrees of freedom, considered in the last chapter where frames close to the mass center, generally, experience large ductility.

Torsional rotation is observed to exist, even at zero eccentricity, in the two-component case (Tables 10.5 to 10.7) as indicated by unequal ductilities in the X and Y-direction ($\overline{D}_{1u} =/= \overline{D}_{4u}, \overline{D}_{2u} =/= \overline{D}_{3u}, \overline{D}_{1v} =/= \overline{D}_{2v}, \overline{D}_{4v} =/= \overline{D}_{3v}$). This reveals the fact that differential yielding in a framed structure causes torsional rotations. This feature was not observed in the structure with two degrees of freedom, dealt with in Chapter 9, because all the frames were assumed to have the same yield displacement. Yielding of the columns is not simultaneous in the present case, because of the consideration of biaxial bending behavior of the columns. In the one-component case, for e/a = 0, torsion is absent, whereas the torsional coupling is reflected by the appearance of Y-components of column ductilities for e/a > 0.

A significant change in the ductility demands of the columns is noticed in the two-component case, as compared with those of the one-component case. Figs. 10.6 to 10.8, which give the ratios of the ductilities in the two-component case to those of one-component case, indicate that eccentricity and period have considerable influence on the change in the ductilities. The ratios of the Y-components of the ductilities are seen to decrease, in general, with increasing eccentricities. The ratio has a maximum

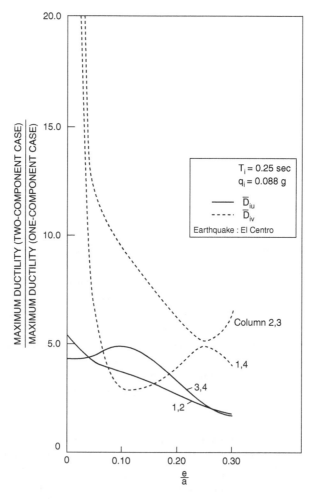

FIGURE 10.6: Variation of the ratio of maximum ductilities (two-component case to one-component case) with eccentricity

Single-Storey Framed Structure: Three Degree of Freedom System

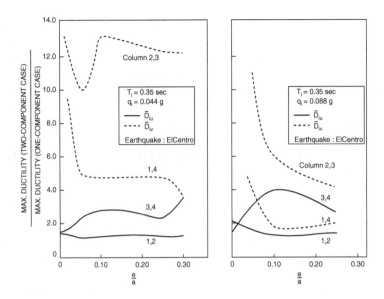

FIGURE 10.7: Variation of the ratio of maximum ductilities (two-component case to one-component case) with eccentricity

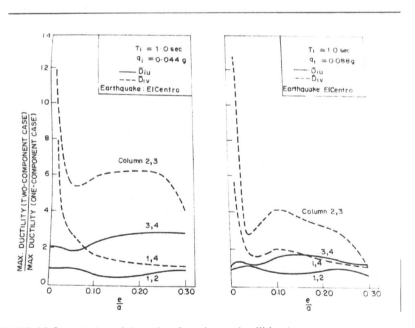

FIGURE 10.8: Variation of the ratio of maximum ductilities (two-component case to one-component case) with eccentricity

value equal to about 5 for the ductilities along the X-direction and can be as high as 13 for the ductilities along the Y-direction. Ductilities are found to be more in the two-component case than those in the one-component case for short-period structures ($T_i = 0.25$ sec, and 0.35 sec). In long-period structures such as $T_i = 1.0$ sec, certain column ductilities ($\overline{D}_{1u}$) are even less, in the two-component case than those of the one-component case. However, the largest column ductility for the two-component case is always greater than the largest column ductility for the one-component case.

10.5.2 INFLUENCE OF YIELD STRENGTH $Q_{IUO} = Q_{IVO} = Q_I$ AND PERIOD ($T_{IU} = T_{IV} = T_I$)

Influence of yield strength may again be seen in Tables 10.3 to 10.7. A decrease in the yield strength from 0.088 g to 0.044 g results in an increase in the value of maximum ductilities, a situation observed in the previous case of a two degrees of freedom system. For instance, in the structure with $T_i = 0.35$ sec., the increase is nearly 100%, in some column ductilities and 200% in some other column ductilities.

Referring to the same tables and the maximum ductility response spectra (Figs. 10.9 and 10.10), it can be noted that structures with columns of short period experience larger ductilities for the El Centro type input. A Similar trend was observed for the structure with two degrees of freedom, considered in the last chapter. Further, the spectra for the two-component case are steeper compared to those of the

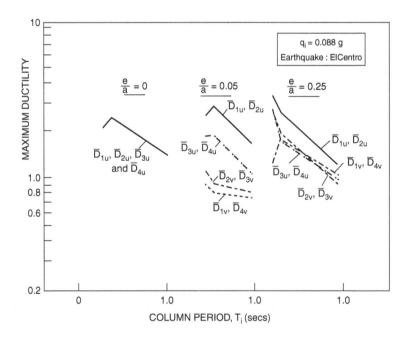

FIGURE 10.9: Column period, $T_i(secs)$

Single-Storey Framed Structure: Three Degree of Freedom System

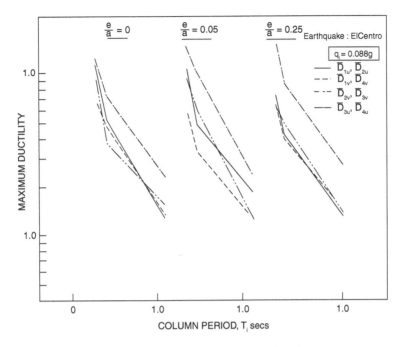

FIGURE 10.10: Column period, $T_i(secs)$

one-component case which also indicates broadly that the increase in the ductility in the two-component case over that in the one-component case is more for short-period structures.

10.5.3 TIME-RESPONSE CURVES

Figs. 10.11 and 10.12 show the response histories for $e/a = 0.0125$ and $e/a = 0.30$, respectively, for the columns 1, 2, 3 and 4. The ground motion input is the El Centro quake. The yield strength of the structure is 0.088 g and its period is 0.35 sec.

The history curves for both the cases are characterized by strong yielding and oscillations around a shifted equilibrium position. In Fig. 10.12 for $e/a = 0.30$, columns 2 and 3 show a mixture of two different kinds of oscillations. Associated with large yielding, the frequency of oscillations is 0.56 Hz, while, when the column is vibrating around a shifted equilibrium position, the frequency of oscillations is 1 Hz. These frequencies may be contrasted with the fundamental linear frequency, which has a value of 2.28 HZ. It is thus clear that the structure is executing, strongly non-linear, oscillations.

Figs. 10.14 and 10.15 show the spatial motion of the center of gravity (mass center) for the structures with $e/a = 0.0125$ and $e/a = 0.30$. It is interesting to note that

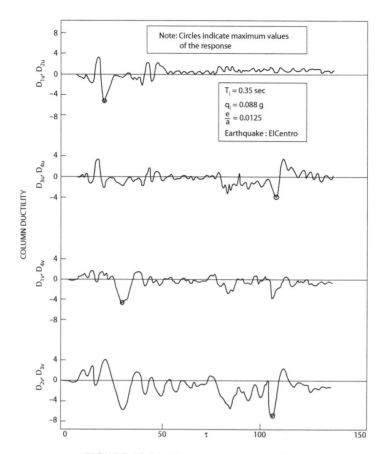

FIGURE 10.11: Time-response relationships

$e/a = 0.30$ the predominant excursions are in the Y-direction. Besides, strong yielding takes place in the Y-direction and once the center of gravity (mass center) moves to one-half of the x-y plane, it remains in that half executing quite a few oscillations around a shifted equilibrium position. During the entire duration of the earthquake the center of gravity crosses the X-axis only four times. A similar trend is seen for $e/a = 0.0125$, although the pre-dominance of the maximum excursion in the Y-direction is not as significant. Also see Fig.10.11. In this case, the mass center crosses the X-axis nine times.

It may, probably, be concluded that the biaxial torsional response displays a direction oriented yielding, wherein the structure yields predominantly in one direction in the X-Y plane depending on the combination of the two ground motion components. Fig. 10.16 shows a more detailed representation of the motion including the rotations of the mass around a vertical axis. The orientations of the axes are shown by a cross, the shorter arms at the right and at the top indicating positive X- and Y-axes

Single-Storey Framed Structure: Three Degree of Freedom System

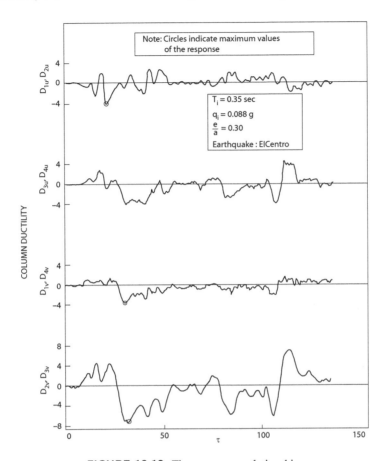

FIGURE 10.12: Time-response relationships

respectively. Time τ corresponding to the various positions of the cross is indicated by the side of the cross. This figure shows response only $\tau = 101$ (corresponding t = 11.22 seconds).

10.6 SUMMARY AND CONCLUSIONS

The inelastic response of a single storeyed structures square in plan and Supported on four columns subjected to ElCentro, May 1940, earthquake has been discussed. The model is assumed to Possess three degrees-of-freedom- Interaction between forces in the two orthogonal directions has been considered. The responses of the structure to simultaneous action of the two Orthogonal, horizontal components of the ground motion and to one of the two components have been compared. The influence of the eccentricity, yield strength and the period of the structure on the response has been studied.

184   Structural Dynamics in Earthquake and Blast Resistant Design

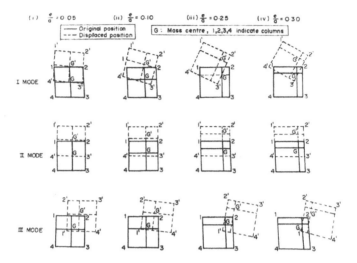

FIGURE 10.13: Mode shapes for various eccentricities

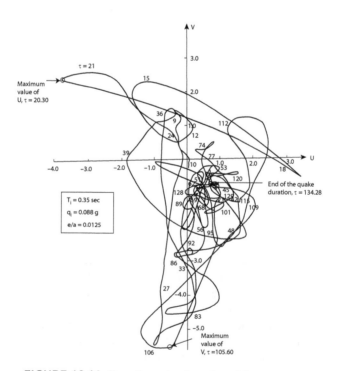

FIGURE 10.14: Two-dimensional motion of the mass center

Single-Storey Framed Structure: Three Degree of Freedom System

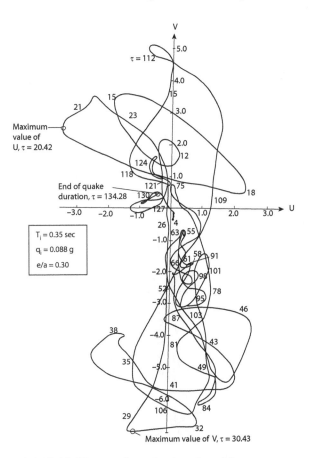

FIGURE 10.15: Two-dimensional motion of the mass center

The investigation showed that torsion does not seem to significantly influence the results for small eccentricities. However, for large eccentricities like e/a = 0.30, although the maximum ductilities of some columns are reduced due to torsion, the largest of the maximum ductilities is increased. This increase ranges from 5% to 50%, when compared with the zero eccentricity case, depending on the period and the yield strength of the structure. It is seen that disparity between the largest of the maximum ductilities ($\overline{D}_{Max}$) to the smallest of the maximum ductilities ($\overline{D}_{Max}$) increases with eccentricity for the two-component input, while it decreases with eccentricity for the one-component input. The columns nearer to the mass center generally experience larger ductilities.

A similar trend was observed for the two degree of freedom system in Chapter 9.

The torsional component of response was noticed even for the zero-eccentricity case for the two-component input. However, the influence of torsion was not seen

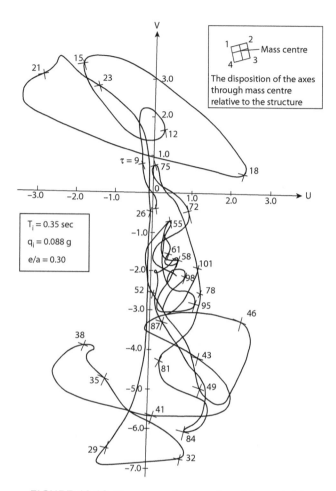

FIGURE 10.16: Two-dimensional motion of the roof slab

for the one-component input, when eccentricity was zero. In the case of the two-component input, yielding is not simultaneous in the columns due to biaxial interaction and eccentricity is introduced in the structure due to differential yielding.

The maximum ductilities are generally larger in the two-component case than those in the one-component case, for short-period structures. For long-period structures, certain column ductilities are smaller in the two-component case. However, the largest of the maximum ductilities is always greater for two-component input.

A reduction in the yield strength by half resulted in a marked increase in the column ductilities, which in some cases can be as large as 200%. The response spectra have shown that columns in short-period structures experience larger ductilities. This feature of inelastic torsional response was noticed in the two degrees of freedom in

Chapter 9. The time-response curves showed that, unlike in the structure with two degrees of freedom, the response behavior could not be explained with reference to linear frequencies and mode shapes. This may be attributed to the strongly non-linear character of the response in the three degrees of freedom model. It was also observed that biaxial torsional response displays a direction oriented yielding, where the structure yields predominantly in one direction in the X-Y plane depending on the combination of the two orthogonal horizontal components of the ground motion.

11 Earthquake Resistant Design as per IS 1893:2016

11.1 INTRODUCTION

In the following chapter we will illustrate several practical examples of fairly tall buildings mostly reinforced concrete constructed in and around Bangalore (Figs. 11.1 through 11.11). The structure designs are prepared by Prasad Consultants, Bangalore for which the founder, CEO is Mr. H.N. Renuka Prasad. Mr. Prasad was author's master's student way back in the late 1980s in the Department of Civil Engineering, Indian Institute of Science, Bangalore. Currently, the author keeps advising him regularly on issues which require special consideration of the IS codes. One such is IS 1893:2016 which has stringent provisions and specifications to be followed during structural design. The important problem here is that while following the architectural consideration it becomes difficult to satisfy IS 1893:2016 while doing earthquake resistant design. Only when some rudimentary concepts of structural dynamics are extended to interpret the IS 1893 is it possible to understand the ramifications of the code. Then with certain structural modifications keeping architectural constraints, it is possible to proceed further satisfying both the architect and IS 1893. With that in view, including such problems in this book has become relevant. We mention a few important issues: avoidance of floating columns, soft storeys, and torsional modes.

All the photographs of the buildings along with the results of the well-known software, ETABS, analysis are taken from Prasad Consultants, Bangalore.

11.2 PROJECT - 01

11.2.1 INTRODUCTION

At the outset, it may not be out of place to outline certain important features of IS 1893 part 1 which will clash with certain architectural features which are being followed almost inevitably because of users' requirements and marketability. For example, use of floating columns has been in vogue for several years. But now the IS 1893:2016 in force specifies such features are undesirable and should be prohibited if they are a part of or supporting primary lateral load resisting system. If the above restriction of the code has to be strictly adhered to, the concept of transfer girders would no longer exist because most of the columns which are supported by the transfer girders would be carrying the major part of the lateral force. The next feature is that of torsional mode. It has been found to be extremely difficult to avoid torsional mode in the first two modes unless there is a thorough revision of all the column stiffnesses to achieve as low as possible the value of eccentricity between center of mass and center of rigidity.

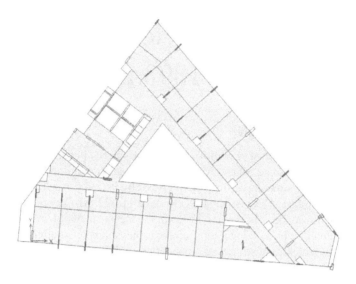

FIGURE 11.1: Project 1. Hotel building structural plan

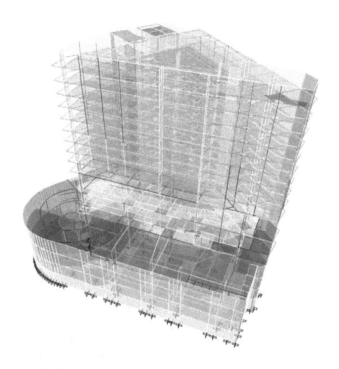

FIGURE 11.2: Project 1. Hotel building analysis model

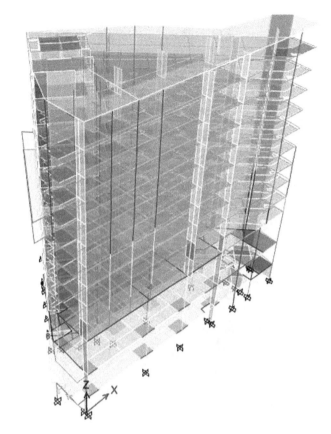

FIGURE 11.3: Project 1.a Hotel building analysis model. fixed at the ground floor level.

The thorough revision of the stiffnesses of the columns calls for a major change in the architecture. So, in the following sections all such features are thoroughly discussed with examples from realistic tall buildings in and around Bangalore.

11.2.2 FLOATING COLUMNS

As discussed earlier, in IS 1893:2016 part I, under 7.1 table 6, it is mentioned that the floating or stub columns cause concentrated damage to the structure and hence are undesirable and should be prohibited if they are part of or supporting lateral load resisting system. So, this became an extremely difficult problem in a hotel project.

The project is a building with four basements, ground floor and eleven upper floors and a terrace situated in Bangalore, which is Zone II. The problem here was that the floating columns are subjected to considerable lateral shear supporting the guest rooms resting on transfer girders.

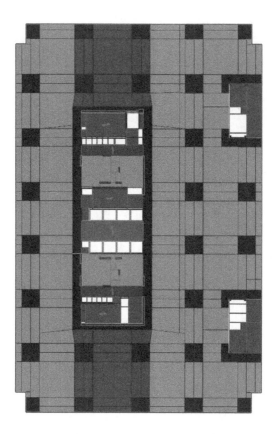

FIGURE 11.4: Project 2. Structural plan (commercial building)

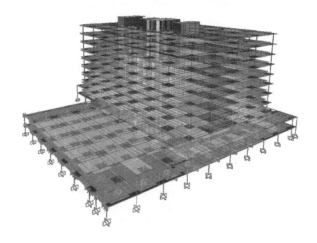

FIGURE 11.5: Project 2. Analysis model

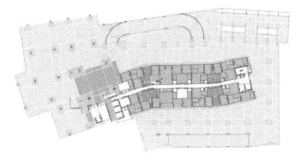

FIGURE 11.6: Project 3. Structural plan (residential and commerical buiding)

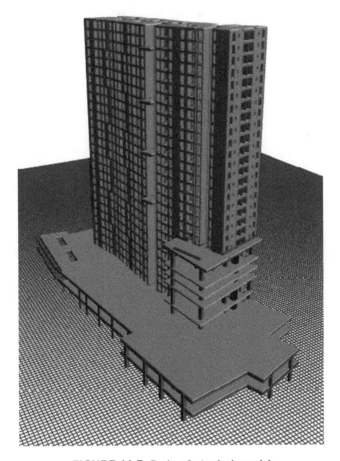

FIGURE 11.7: Project 3. Analysis model

194 Structural Dynamics in Earthquake and Blast Resistant Design

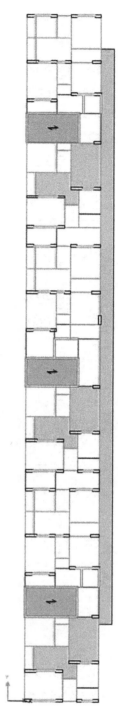

FIGURE 11.8: Project 4. Structural plan (without structural joint)

Earthquake Resistant Design as per IS 1893:2016 195

FIGURE 11.9: Project 4. Analysis model (without structural joint)

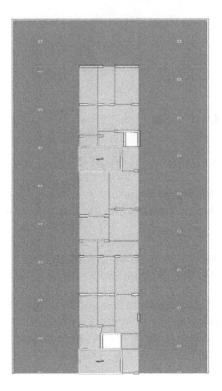

FIGURE 11.10: Project 4. Structural plan (with structural joint)

FIGURE 11.11: Project 4. Analysis model (with structural joint)

In order to comply with the requirements of IS 1893, a few floating columns were released from carrying the lateral shear for which the lateral stiffness of those columns was reduced significantly. The above modification reduced lateral shear significantly in those floating columns, but because of the low stiffness it resulted in creating a soft storey at that level.

It means that in trying to circumvent the floating columns from carrying the lateral shear, we ended up with a problem of soft storey which is not allowed by IS 1893:2016.

The IS 1893 states that the lateral stiffness of the storey below should not be less that of the above, or in other words a soft storey should not exist in a building if it has to be considered as a building without vertical irregularity.

11.2.3 SOFT STOREY

As said above in the preceding section in trying to avoid lateral shear being transferred to floating columns, the stiffness of those floating columns is reduced accordingly which resulted in soft storey. Again, as understood earlier, IS 1893 does not allow soft storey. Then the question is how to circumvent both of the above problems, viz., that of floating columns and that of soft storey. The solution we envisaged in the above project was to take care of both. One of the ways is to adjust the stiffnesses of the upper storey to be lower than that of the storey containing floating columns and finally to transfer balance lateral shear which resulted in reducing stiffnesses of all the storeys from the storey containing floating columns till the top, to a well-designed shear wall in the building. In fact, clause 7.3, table 9, of IS 1893:2016 under flat slab specifies that ductile RC structural walls (shear walls) to be designed to resist 100% lateral force. It was invoked here also, although in IS 1893 it is mentioned only under flat slabs.

11.2.4 BUILDING ASYMMETRIC IN PLAN

The buildings which are asymmetric in plan, i.e., when there is eccentricity between the center of mass and center of rigidity (center of stiffness), are again very specially discussed in IS 1893. The problem in such buildings is that because of eccentricity, the building rotates in plan during translational motions although ground motion is only translational. Although it is not possible to avoid torsion completely, it can be minimized by reducing the eccentricity which is affected by adjusting the lateral stiffnesses of various columns and shear walls. In order to ascertain that torsion has been minimized, we check whether the first two modes do not have predominant torsion component. IS 1893 also specifies that a building is said to be torsionally irregular when the maximum horizontal displacement of any floor in the direction of the lateral force at one end of the floor is more than 1.5 times its minimum horizontal displacement at the far end of the same floor in that direction, and the natural period corresponding to the fundamental torsional mode of oscillation is less than those of the first two translational modes of oscillation along each principal plan direction.

In most of the projects, it was always very difficult to achieve lowest torsional component compared to other two translational components in the first two modes. The results of the ETABS regarding the modal components in different modes are shown in tables 1 to 13 for the respective projects 1, 2, 3 and 4. It can be seen from the table that although the torsional component is much more than the other two components in the first few modes, the torsional component is highest in the 16th mode in project 1, 18th mode in project 2, 22nd mode in project 3, and 9th mode in project 4. The translational modal component in X direction is largest in mode 3 for project 2 and mode 1 for project 4. Similarly, translations in Y direction are largest in mode 2 for project 1, mode 4 for project 2, in mode 1 for project 3 and in mode 3 for project 4.

Therefore, it was suggested that it would be sufficient that predominant torsional mode has a period much lower than the first two translational components although they are not strictly the first three modes. Perhaps this could be a modification to be suggested to IS 1893 when it is extremely difficult to achieve least torsion in the first two modes strictly. Another point worth mentioning here is that when looking for a largest torsional component in one of the higher modes as observed above, that mode should be within the total number of modes chosen to satisfy the condition that cumulative mass participation factor up to that mode is around 65%.

Table 01: Project 1 Modal Direction Factors

Case	Mode	Period	UX	UY	RZ
		sec			
Modal	1	2.710	0.367	0.190	0.442
Modal	2	2.009	0.051	0.809	0.140
Modal	3	1.831	0.582	0.012	0.406
Modal	4	0.800	0.339	0.179	0.482
Modal	5	0.604	0.002	0.264	0.734
Modal	6	0.505	0.005	0.636	0.360
Modal	7	0.444	0.558	0.130	0.312
Modal	8	0.430	0.052	0.013	0.935
Modal	9	0.408	0.257	0.079	0.664
Modal	10	0.393	0.011	0.011	0.978
Modal	11	0.393	0.009	0.013	0.978
Modal	12	0.393	0.007	0.012	0.981
Modal	13	0.393	0.007	0.012	0.981
Modal	14	0.393	0.007	0.012	0.982
Modal	15	0.393	0.007	0.012	0.981
Modal	16	0.393	0.006	0.011	0.983
Modal	17	0.392	0.011	0.014	0.975
Modal	18	0.385	0.005	0.061	0.934
Modal	19	0.382	0.382	0.407	0.212
Modal	20	0.231	0.000	0.003	0.997
Modal	21	0.217	0.091	0.727	0.182
Modal	22	0.215	0.395	0.307	0.299
Modal	23	0.200	0.552	0.060	0.389
Modal	24	0.199	0.565	0.056	0.379
Modal	25	0.192	0.427	0.184	0.388
Modal	26	0.189	0.457	0.180	0.362
Modal	27	0.161	0.001	0.009	0.989
Modal	28	0.146	0.485	0.391	0.123
Modal	29	0.135	0.296	0.321	0.382
Modal	30	0.127	0.352	0.441	0.206

Table 02: Project 1 Modal Participating Mass Ratios

Mode	Period sec	UX	UY	Sum UX	Sum UY	RX	RY	RZ	Sum RX	Sum RY	Sum RZ
1	2.710	0.120	0.060	0.120	0.060	0.111	0.222	0.101	0.111	0.222	0.101
2	2.009	0.017	0.277	0.137	0.337	0.486	0.031	0.008	0.597	0.253	0.109
3	1.831	0.212	0.001	0.348	0.338	0.004	0.348	0.095	0.601	0.600	0.204
4	0.800	0.033	0.026	0.381	0.364	0.000	0.001	0.045	0.601	0.602	0.249
5	0.604	0.000	0.001	0.381	0.364	0.000	0.000	0.000	0.601	0.602	0.249
6	0.505	0.001	0.087	0.382	0.451	0.002	0.000	0.008	0.603	0.602	0.257
7	0.444	0.088	0.007	0.470	0.458	0.000	0.002	0.007	0.603	0.603	0.263
8	0.430	0.002	0.000	0.472	0.458	0.000	0.000	0.002	0.603	0.603	0.265
9	0.408	0.004	0.008	0.476	0.466	0.000	0.000	0.020	0.603	0.603	0.285
10	0.393	0.000	0.000	0.476	0.466	0.000	0.000	0.000	0.603	0.603	0.286
11	0.393	0.000	0.000	0.476	0.466	0.000	0.000	0.000	0.603	0.603	0.286
12	0.393	0.000	0.000	0.476	0.466	0.000	0.000	0.000	0.603	0.603	0.286
13	0.393	0.000	0.000	0.476	0.466	0.000	0.000	0.000	0.603	0.603	0.286
14	0.393	0.000	0.000	0.476	0.466	0.000	0.000	0.000	0.603	0.603	0.286
15	0.393	0.000	0.000	0.476	0.466	0.000	0.000	0.000	0.603	0.603	0.286
16	0.393	0.000	0.000	0.476	0.466	0.000	0.000	0.000	0.603	0.603	0.286
17	0.392	0.000	0.000	0.476	0.466	0.000	0.000	0.000	0.603	0.603	0.286
18	0.385	0.000	0.002	0.476	0.469	0.000	0.000	0.000	0.604	0.603	0.286
19	0.382	0.000	0.004	0.476	0.472	0.000	0.000	0.007	0.604	0.604	0.293
20	0.231	0.000	0.000	0.476	0.472	0.000	0.000	0.000	0.604	0.604	0.293
21	0.217	0.002	0.046	0.478	0.519	0.008	0.001	0.000	0.612	0.604	0.294
22	0.215	0.009	0.003	0.487	0.522	0.000	0.002	0.002	0.612	0.606	0.295
23	0.200	0.007	0.001	0.495	0.522	0.000	0.001	0.003	0.613	0.607	0.298
24	0.199	0.028	0.002	0.523	0.525	0.001	0.005	0.007	0.614	0.612	0.305
25	0.192	0.000	0.000	0.523	0.525	0.000	0.000	0.000	0.614	0.612	0.305
26	0.189	0.000	0.000	0.523	0.525	0.000	0.000	0.000	0.614	0.612	0.305
27	0.161	0.000	0.002	0.523	0.527	0.001	0.000	0.000	0.615	0.613	0.305
28	0.146	0.079	0.122	0.602	0.649	0.050	0.035	0.000	0.665	0.647	0.306
29	0.135	0.008	0.064	0.610	0.713	0.030	0.004	0.009	0.694	0.652	0.315
30	0.127	0.041	0.048	0.650	0.761	0.020	0.014	0.004	0.714	0.665	0.319

Table 03: Project 1.a Modal Participating Mass Ratios											
	Period sec	UX	UY	Sum UX	Sum UY	RX	RY	RZ	Sum RX	Sum RY	Sum RZ
1	2.571	0.198	0.109	0.198	0.109	0.068	0.127	0.298	0.068	0.127	0.298
2	1.825	0.003	0.487	0.201	0.596	0.291	0.002	0.153	0.360	0.129	0.451
3	1.677	0.438	0.027	0.639	0.623	0.021	0.238	0.159	0.380	0.367	0.610
4	0.777	0.060	0.045	0.699	0.668	0.037	0.091	0.119	0.417	0.457	0.729
5	0.462	0.000	0.141	0.699	0.809	0.184	0.001	0.039	0.601	0.458	0.768
6	0.430	0.002	0.007	0.702	0.816	0.010	0.003	0.005	0.610	0.462	0.772
7	0.416	0.094	0.001	0.796	0.817	0.001	0.128	0.000	0.611	0.589	0.772
8	0.400	0.070	0.018	0.866	0.834	0.018	0.084	0.052	0.629	0.674	0.824
9	0.393	0.000	0.000	0.866	0.834	0.000	0.000	0.001	0.629	0.674	0.825
10	0.393	0.000	0.000	0.866	0.834	0.000	0.000	0.000	0.629	0.674	0.825
11	0.393	0.000	0.000	0.866	0.834	0.000	0.000	0.000	0.629	0.674	0.825
12	0.393	0.000	0.000	0.866	0.834	0.000	0.000	0.000	0.629	0.674	0.825
13	0.393	0.000	0.000	0.866	0.835	0.000	0.000	0.000	0.629	0.674	0.825
14	0.393	0.000	0.000	0.866	0.835	0.000	0.000	0.000	0.629	0.674	0.825
15	0.393	0.000	0.000	0.866	0.835	0.000	0.000	0.000	0.629	0.674	0.825
16	0.392	0.000	0.000	0.866	0.835	0.001	0.000	0.000	0.630	0.674	0.825
17	0.383	0.001	0.010	0.867	0.845	0.013	0.001	0.004	0.642	0.675	0.829
18	0.380	0.000	0.009	0.867	0.854	0.011	0.000	0.023	0.653	0.675	0.852
19	0.213	0.004	0.001	0.871	0.856	0.004	0.011	0.004	0.657	0.686	0.856
20	0.206	0.007	0.033	0.878	0.889	0.060	0.013	0.003	0.717	0.698	0.859
21	0.200	0.000	0.000	0.878	0.889	0.001	0.000	0.000	0.717	0.699	0.859
22	0.192	0.000	0.000	0.878	0.890	0.001	0.000	0.001	0.718	0.699	0.860
23	0.189	0.001	0.000	0.879	0.890	0.001	0.002	0.001	0.718	0.700	0.861
24	0.188	0.022	0.003	0.901	0.893	0.006	0.040	0.020	0.724	0.740	0.881
25	0.136	0.003	0.001	0.904	0.893	0.002	0.008	0.006	0.726	0.748	0.887
26	0.121	0.013	0.002	0.917	0.895	0.005	0.032	0.005	0.731	0.780	0.892
27	0.113	0.009	0.008	0.925	0.903	0.017	0.024	0.009	0.748	0.804	0.902
28	0.102	0.002	0.005	0.927	0.908	0.013	0.004	0.009	0.760	0.807	0.911
29	0.096	0.003	0.000	0.930	0.908	0.000	0.007	0.000	0.761	0.814	0.911
30	0.084	0.001	0.005	0.930	0.913	0.013	0.001	0.000	0.774	0.815	0.912

Table 04: Project 2 Modal Direction Factors					
Case	Mode	Period	UX	UY	RZ
		sec			
Modal	1	1.600	0.009	0.911	0.08
Modal	2	1.528	0.002	0.082	0.916
Modal	3	1.128	0.989	0.010	0.000
Modal	4	0.477	0.002	0.986	0.012
Modal	5	0.397	0.000	1.000	0.000
Modal	6	0.395	0.001	0.017	0.981
Modal	7	0.294	0.049	0.652	0.299
Modal	8	0.288	1.000	0.000	0.000
Modal	9	0.267	0.971	0.024	0.005
Modal	10	0.264	0.000	0.000	1.000
Modal	11	0.236	0.019	0.762	0.218
Modal	12	0.205	0.158	0.716	0.125
Modal	13	0.198	0.634	0.252	0.114
Modal	14	0.194	0.003	0.017	0.979
Modal	15	0.181	0.025	0.166	0.809
Modal	16	0.163	0.035	0.563	0.402
Modal	17	0.143	0.128	0.052	0.820
Modal	18	0.133	0.001	0.008	0.992
Modal	19	0.127	0.831	0.041	0.128
Modal	20	0.122	0.061	0.754	0.185
Modal	21	0.121	0.663	0.158	0.179
Modal	22	0.115	0.342	0.004	0.654
Modal	23	0.113	0.327	0.019	0.655
Modal	24	0.111	0.194	0.004	0.802
Modal	25	0.106	0.216	0.010	0.775
Modal	26	0.098	0.033	0.548	0.419
Modal	27	0.098	0.023	0.197	0.780
Modal	28	0.093	0.000	0.042	0.958
Modal	29	0.088	0.001	0.630	0.369
Modal	30	0.086	0.014	0.340	0.647

Table 05: Project 2 Modal Participating Mass Ratios

Mode	Period sec	UX	UY	Sum UX	Sum UY	RX	RY	RZ	Sum RX	Sum RY	Sum RZ
1	1.6	0.006	0.620	0.006	0.620	0.294	0.003	0.058	0.294	0.003	0.058
2	1.528	0.001	0.051	0.006	0.671	0.029	0.001	0.496	0.323	0.004	0.554
3	1.128	0.589	0.007	0.596	0.679	0.003	0.406	0.001	0.326	0.410	0.554
4	0.477	0.000	0.128	0.596	0.807	0.308	0.000	0.001	0.634	0.410	0.555
5	0.397	0.000	0.002	0.596	0.809	0.003	0.000	0.004	0.637	0.410	0.559
6	0.395	0.000	0.005	0.596	0.813	0.006	0.000	0.157	0.643	0.410	0.716
7	0.294	0.009	0.062	0.605	0.875	0.101	0.012	0.070	0.744	0.422	0.786
8	0.288	0.002	0.000	0.607	0.875	0.000	0.003	0.000	0.744	0.425	0.787
9	0.267	0.168	0.001	0.775	0.876	0.002	0.183	0.002	0.746	0.608	0.788
10	0.264	0.000	0.000	0.775	0.876	0.000	0.000	0.000	0.746	0.608	0.788
11	0.236	0.001	0.033	0.776	0.909	0.048	0.002	0.044	0.794	0.610	0.833
12	0.205	0.005	0.003	0.781	0.912	0.006	0.008	0.002	0.799	0.618	0.835
13	0.198	0.034	0.000	0.815	0.912	0.001	0.056	0.000	0.800	0.674	0.835
14	0.194	0.001	0.001	0.816	0.913	0.002	0.001	0.002	0.802	0.675	0.837
15	0.181	0.004	0.003	0.820	0.916	0.007	0.006	0.028	0.808	0.681	0.865
16	0.163	0.005	0.026	0.824	0.942	0.056	0.007	0.020	0.864	0.688	0.885
17	0.143	0.014	0.001	0.838	0.944	0.003	0.023	0.011	0.867	0.711	0.896
18	0.133	0.000	0.000	0.838	0.944	0.000	0.000	0.001	0.867	0.711	0.897
19	0.127	0.021	0.000	0.859	0.944	0.000	0.031	0.002	0.867	0.742	0.898
20	0.122	0.003	0.001	0.862	0.945	0.002	0.004	0.003	0.869	0.746	0.902
21	0.121	0.011	0.000	0.873	0.945	0.000	0.014	0.000	0.869	0.760	0.902
22	0.115	0.002	0.000	0.875	0.945	0.000	0.003	0.002	0.869	0.763	0.904
23	0.113	0.000	0.000	0.875	0.945	0.001	0.000	0.002	0.870	0.763	0.906
24	0.111	0.007	0.000	0.882	0.945	0.000	0.012	0.000	0.870	0.775	0.906
25	0.106	0.001	0.000	0.883	0.945	0.000	0.002	0.000	0.870	0.777	0.906
26	0.098	0.000	0.001	0.883	0.946	0.001	0.001	0.001	0.871	0.777	0.907
27	0.098	0.000	0.002	0.883	0.948	0.004	0.001	0.002	0.875	0.778	0.910
28	0.093	0.000	0.000	0.883	0.948	0.000	0.000	0.000	0.875	0.778	0.910
29	0.088	0.000	0.000	0.883	0.948	0.000	0.000	0.000	0.875	0.778	0.910
30	0.086	0.000	0.011	0.883	0.959	0.027	0.000	0.026	0.901	0.778	0.936

Table 06: Project 3 Modal Direction Factors

Case	Mode	Period	UX	UY	RZ
Modal	1	1.513	0.001	**0.983**	0.015
Modal	2	1.008	0.066	0.023	0.911
Modal	3	0.887	**0.931**	0.001	0.067
Modal	4	0.391	0.009	0.687	0.304
Modal	5	0.310	0.110	0.338	0.552
Modal	6	0.275	0.811	0.025	0.164
Modal	7	0.211	0.173	0.040	0.787
Modal	8	0.199	0.167	0.020	0.814
Modal	9	0.177	0.293	0.095	0.612
Modal	10	0.165	0.601	0.262	0.137
Modal	11	0.159	0.028	0.936	0.035
Modal	12	0.147	0.011	0.000	0.989
Modal	13	0.134	0.036	0.073	0.891
Modal	14	0.116	0.860	0.009	0.132
Modal	15	0.107	0.038	0.885	0.077
Modal	16	0.106	0.088	0.448	0.464
Modal	17	0.102	0.228	0.355	0.417
Modal	18	0.100	0.037	0.404	0.559
Modal	19	0.092	0.204	0.043	0.752
Modal	20	0.090	0.512	0.012	0.476
Modal	21	0.087	0.109	0.056	0.834
Modal	22	0.084	0.002	0.006	**0.992**
Modal	23	0.076	0.034	0.066	0.900
Modal	24	0.076	0.025	0.051	0.924
Modal	25	0.073	0.210	0.326	0.464

Table 07: Project 3 Modal Participating Mass Ratios

Case	Mode	Period	UX	UY	Sum UX	Sum UY	RX	RY	RZ	Sum RX	Sum RY	Sum RZ
Modal	1	1.513	0.001	0.545	0.001	0.545	0.451	0.001	0.018	0.451	0.001	0.018
Modal	2	1.008	0.040	0.005	0.040	0.550	0.012	0.032	0.409	0.463	0.033	0.427
Modal	3	0.887	0.554	0.000	0.595	0.550	0.000	0.453	0.037	0.463	0.486	0.464
Modal	4	0.391	0.003	0.114	0.598	0.663	0.120	0.001	0.035	0.584	0.487	0.498
Modal	5	0.310	0.028	0.060	0.627	0.723	0.048	0.035	0.097	0.632	0.522	0.595
Modal	6	0.275	0.193	0.002	0.819	0.725	0.001	0.248	0.029	0.633	0.771	0.624
Modal	7	0.211	0.003	0.000	0.822	0.725	0.000	0.005	0.074	0.633	0.776	0.698
Modal	8	0.199	0.000	0.000	0.823	0.726	0.001	0.004	0.098	0.634	0.780	0.796
Modal	9	0.177	0.004	0.002	0.827	0.728	0.003	0.014	0.057	0.637	0.794	0.852
Modal	10	0.165	0.001	0.010	0.828	0.738	0.014	0.007	0.018	0.651	0.801	0.870
Modal	11	0.159	0.000	0.102	0.828	0.840	0.120	0.000	0.007	0.771	0.801	0.878
Modal	12	0.147	0.000	0.000	0.828	0.840	0.000	0.000	0.000	0.771	0.801	0.878
Modal	13	0.134	0.000	0.013	0.828	0.853	0.013	0.000	0.011	0.785	0.802	0.888
Modal	14	0.116	0.001	0.001	0.829	0.854	0.001	0.000	0.003	0.786	0.802	0.891
Modal	15	0.107	0.003	0.051	0.832	0.905	0.071	0.003	0.000	0.857	0.805	0.891
Modal	16	0.106	0.002	0.001	0.834	0.906	0.002	0.002	0.000	0.858	0.806	0.891
Modal	17	0.102	0.015	0.022	0.848	0.928	0.031	0.014	0.007	0.889	0.820	0.898
Modal	18	0.100	0.000	0.001	0.848	0.929	0.001	0.000	0.000	0.891	0.821	0.898
Modal	19	0.092	0.003	0.002	0.851	0.931	0.003	0.006	0.022	0.894	0.827	0.920
Modal	20	0.090	0.000	0.001	0.851	0.932	0.002	0.000	0.000	0.895	0.827	0.921
Modal	21	0.087	0.019	0.006	0.870	0.938	0.009	0.018	0.003	0.904	0.845	0.924
Modal	22	0.084	0.000	0.001	0.870	0.939	0.002	0.000	0.000	0.906	0.845	0.924
Modal	23	0.076	0.000	0.003	0.870	0.943	0.005	0.000	0.000	0.911	0.845	0.924
Modal	24	0.076	0.000	0.001	0.870	0.944	0.002	0.000	0.000	0.913	0.845	0.925
Modal	25	0.073	0.000	0.006	0.870	0.950	0.009	0.000	0.004	0.922	0.845	0.928

TABLE 08: Project 4 Modal Direction Factors

Case	Mode	Period	UX	UY	RZ
Modal	1	0.873	0.982	0.001	0.017
Modal	2	0.706	0.018	0.013	0.969
Modal	3	0.496	0.000	0.987	0.013
Modal	4	0.163	0.759	0.104	0.136
Modal	5	0.158	0.159	0.819	0.022
Modal	6	0.151	0.082	0.078	0.840
Modal	7	0.080	0.013	0.968	0.018
Modal	8	0.073	0.775	0.025	0.201
Modal	9	0.070	0.155	0.005	0.841
Modal	10	0.051	0.033	0.884	0.083
Modal	11	0.048	0.525	0.038	0.437
Modal	12	0.046	0.073	0.002	0.925

Earthquake Resistant Design as per IS 1893:2016

TABLE 09: Project 4 Modal Participating Mass Ratios

Case	Mode	Period	UX	UY	Sum UX	Sum UY	RX	RY	RZ	Sum RX	Sum RY	Sum RZ
Modal	1	0.873	0.584	0.001	0.584	0.001	0.000	0.432	0.006	0.000	0.432	0.006
Modal	2	0.706	0.011	0.008	0.595	0.009	0.005	0.008	0.569	0.006	0.439	0.575
Modal	3	0.496	0.000	0.729	0.595	0.738	0.281	0.000	0.008	0.287	0.439	0.583
Modal	4	0.163	0.153	0.010	0.748	0.747	0.041	0.155	0.022	0.328	0.594	0.605
Modal	5	0.158	0.033	0.078	0.781	0.825	0.343	0.035	0.007	0.671	0.629	0.611
Modal	6	0.151	0.017	0.007	0.798	0.833	0.035	0.019	0.149	0.706	0.648	0.760
Modal	7	0.080	0.002	0.022	0.800	0.855	0.038	0.002	0.002	0.743	0.650	0.763
Modal	8	0.073	0.077	0.001	0.877	0.856	0.002	0.121	0.013	0.745	0.771	0.775
Modal	9	0.070	0.017	0.000	0.894	0.856	0.000	0.028	0.069	0.745	0.798	0.844
Modal	10	0.051	0.003	0.038	0.897	0.894	0.087	0.006	0.002	0.832	0.804	0.846
Modal	11	0.048	0.053	0.005	0.950	0.899	0.010	0.098	0.013	0.841	0.902	0.859
Modal	12	0.046	0.011	0.001	0.961	0.900	0.001	0.021	0.051	0.843	0.923	0.910

TABLE 10: Project 4, Modal Direction Factors

Case	Mode	Period	UX	UY	UZ	RZ
Modal	1	0.946	**0.999**	0.000	0.000	0.000
Modal	2	0.617	0.008	0.151	0.000	0.841
Modal	3	0.511	0.001	0.853	0.000	0.146
Modal	4	0.175	0.986	0.000	0.000	0.014
Modal	5	0.165	0.001	0.938	0.000	0.060
Modal	6	0.150	0.024	0.070	0.000	0.906
Modal	7	0.084	0.039	**0.949**	0.000	0.012
Modal	8	0.081	0.932	0.034	0.000	0.035
Modal	9	0.072	0.045	0.016	0.000	**0.938**
Modal	10	0.056	0.004	0.674	0.000	0.322
Modal	11	0.056	0.006	0.465	0.000	0.529
Modal	12	0.055	0.047	0.921	0.000	0.032

TABLE 11: Project 4 Modal Participating Mass Ratios

Case	Mode	Period	UX	UY	Sum UX	Sum UY	RX	RY	RZ	Sum RX	Sum RY	Sum RZ
Modal	1	0.946	0.586	0.000	0.586	0.000	0.000	0.440	0.003	0.000	0.440	0.003
Modal	2	0.617	0.000	0.096	0.587	0.096	0.058	0.000	0.454	0.058	0.440	0.457
Modal	3	0.511	0.000	0.588	0.587	0.683	0.283	0.000	0.083	0.342	0.440	0.539
Modal	4	0.175	0.220	0.000	0.806	0.683	0.000	0.228	0.005	0.342	0.668	0.545
Modal	5	0.165	0.000	0.111	0.806	0.794	0.312	0.000	0.013	0.654	0.668	0.558
Modal	6	0.150	0.002	0.008	0.808	0.802	0.027	0.001	0.134	0.681	0.669	0.692
Modal	7	0.084	0.005	0.034	0.813	0.836	0.049	0.007	0.001	0.730	0.676	0.693
Modal	8	0.081	0.101	0.001	0.914	0.837	0.002	0.164	0.004	0.732	0.839	0.697
Modal	9	0.072	0.002	0.000	0.917	0.837	0.001	0.003	0.083	0.733	0.842	0.779
Modal	10	0.056	0.000	0.015	0.917	0.852	0.029	0.000	0.001	0.762	0.842	0.780
Modal	11	0.056	0.000	0.013	0.917	0.865	0.025	0.000	0.004	0.787	0.842	0.784
Modal	12	0.055	0.003	0.030	0.920	0.895	0.056	0.005	0.001	0.843	0.847	0.785

11.2.5 MASS PARTICIPATION FACTOR

An important question in dynamic analysis is how many modes have to be considered in capturing the actual response of the structure. Very often in the early days it is mentioned fequently in all literature that the first three should be sufficient enough for all buildings considered as shear buildings with only translational degrees of freedom in the direction of earthquake. However, with the advent of modern softwares, like ETABS, ROBOT, SAP, GT STRUDL, MIDAS, STAAD PRO, buildings are very conveniently analyzed as space frames with six degrees of freedom per node. With so many degrees of freedom considered in the building with numerous degrees of freedom, the first three modes would not be sufficient for mass participation. The adequate mass participation is 65% of the total seismic mass. Therefore, the total number of modes required to achieve an accurate response is mass participation of 65%.

In a few projects it has become quite challenging to achieve the required mass participation factor without certain modifications. For example, in project 1 shown in Fig. 11.3 the structure was considered to be fixed against rotations at the base below the four basements. The basement floors were as usual surrounded by the retaining walls which were also a part of the structure which meant additional lateral constraint from the sides. The mass participation factor even after several modes could not sum up to 65%. Then it was thought that because the structure is fixed at the base and also because the structure is constrained by the retaining walls, the relative ordinates of the eigenvector up to the podium level would be quite less, and therefore if the structure is considered in two parts, viz. one above the podium consisting of all the upper storeys and below consisting of four basements, it could give better values.

In project 1, The building has 11 upper floors and a terrace above the podium. One slight modification was to treat the structure in two parts, one above the podium consisting all the upper floors and one part below the podium consisting of two basements. The basements are surrounded by the retaining walls on all the sides. The columns of the upper part were assumed to be fixed against rotation at the podium level and the part below was assumed to be fixed against rotation at the base. The two parts were analyzed separately, and it was possible to obtain around 65% mass participation factor within first three modes for the upper part. The reason apparently could be the columns of the upper floors being fixed against rotations at the podium level do not have lateral constraint from the surrounding retaining walls, and therefore the amplitudes of the eigenvectors will have relatively larger values compared to the structure which had fixity at the bottom of the two basements. The basement structure is separately analyzed and even that also had a good mass participation factor. Next, one check was run to ensure that the assumption of fixity at the podium level was justified. It was done by ensuring that the rotations are negligibly small [refer to the following Tables 12 and 13] at the podium level compared to those at the top of the upper part. Similarly, a small modification in a very long narrow building with a long aspect ratio in order to avoid torsional component in modes of larger period, the building was divided in two halves with an expansion joint in the middle. In project 4, before providing a structural joint, the fundamental torsional mode 2 (Table 8), and fundamental translational modes in 1 and 3, are X- and Y-directions, respectively. Hence torsional irregularity exits in the structure. After providing the structural joint, fundamental torsional mode shifted to 9th mode (Table 10), as fundamental translational mode in 1 and 7 are X- and Y-directions, respectively, i.e., torsional irregularity does not exist in this structure after providing structural joint.

TABLE 12: Joint Displacements

Storey	Label	Load Case/Combo	UX	UY	RX	RY	RZ
			mm	mm	rad	rad	rad
GROUND FLOOR	174	DL+LL+RSX Max	0.013000	-0.059000	0.000072	-0.000015	0.000003
GROUND FLOOR	2	DL+LL+RSX Max	0.017000	-0.045000	0.000078	-0.000023	0.000003
GROUND FLOOR	3	DL+LL+RSX Max	0.021000	-0.047000	0.000149	-0.000084	0.000003
GROUND FLOOR	4	DL+LL+RSX Max	0.071000	-0.009000	0.000101	-0.000011	0.000002
GROUND FLOOR	5	DL+LL+RSX Max	0.030000	-0.032000	0.000112	-0.000019	0.000003
GROUND FLOOR	6	DL+LL+RSX Max	0.031000	-0.010000	0.000119	0.000002	0.000002
GROUND FLOOR	28	DL+LL+RSX Max	0.025000	-0.032000	0.000052	0.000037	0.000000
GROUND FLOOR	19	DL+LL+RSX Max	0.038000	-0.081000	0.000136	0.000067	0.000003
GROUND FLOOR	20	DL+LL+RSX Max	0.040000	-0.046000	0.001947	-0.000373	0.000003
GROUND FLOOR	21	DL+LL+RSX Max	0.039000	-0.081000	0.000290	0.000430	0.000003
GROUND FLOOR	27	DL+LL+RSX Max	0.041000	-0.046000	-0.000398	-0.000368	0.000003
GROUND FLOOR	29	DL+LL+RSX Max	0.042000	-0.021000	0.000004	0.000055	0.000003
GROUND FLOOR	64	DL+LL+RSX Max	0.041000	-0.038000	0.000057	-0.005376	0.000003
GROUND FLOOR	87	DL+LL+RSX Max	0.034000	-0.083000	0.000303	-0.000410	0.000003
GROUND FLOOR	42	DL+LL+RSX Max	0.021000	-0.067000	0.000088	0.000163	0.000003
GROUND FLOOR	63	DL+LL+RSX Max	0.036000	-0.093000	0.000088	-0.000026	0.000015
GROUND FLOOR	91	DL+LL+RSX Max	0.021000	-0.068000	-0.000679	-0.000364	0.000003
GROUND FLOOR	96	DL+LL+RSX Max	0.029000	-0.003000	0.000107	0.000031	0.000004
GROUND FLOOR	152	DL+LL+RSX Max	0.025000	-0.073000	-0.000896	-0.000849	0.000003
GROUND FLOOR	143	DL+LL+RSX Max	0.023000	-0.036000	0.000019	-0.000013	0.000021
GROUND FLOOR	67	DL+LL+RSX Max	0.034000	-0.025000	0.000114	0.000010	0.000003
GROUND FLOOR	218	DL+LL+RSX Max	0.018000	-0.065000	0.000057	-0.000181	0.000003
GROUND FLOOR	219	DL+LL+RSX Max	0.017000	-0.064000	0.000234	-0.000017	0.000003
GROUND FLOOR	220	DL+LL+RSX Max	0.019000	-0.066000	-0.000245	-0.000740	0.000003
GROUND FLOOR	283	DL+LL+RSX Max	0.040000	-0.067000	0.000315	0.001082	0.000003
GROUND FLOOR	285	DL+LL+RSX Max	0.040000	-0.057000	-0.000089	0.001741	0.000003
GROUND FLOOR	287	DL+LL+RSX Max	0.018000	-0.065000	-0.000040	-0.000278	0.000003
GROUND FLOOR	305	DL+LL+RSX Max	0.041000	-0.031000	0.000230	-0.002108	0.000003
GROUND FLOOR	373	DL+LL+RSX Max	0.104000	0.029000	0.000069	0.000019	0.000041
GROUND FLOOR	1024	DL+LL+RSX Max	0.026000	-0.074000	-0.000772	-0.000776	0.000003
GROUND FLOOR	1003	DL+LL+RSX Max	0.026000	-0.074000	-0.000802	-0.000816	0.000003

TABLE 13: Joint Displacements

Storey	Label	Load Case/Combo	UX	UY	RX	RY	RZ
			mm	mm	rad	rad	rad
GROUND FLOOR	174	DL+LL+RSY Max	-0.028000	0.042000	0.000096	0.000001	0.000003
GROUND FLOOR	2	DL+LL+RSY Max	-0.023000	0.055000	0.000098	-0.000013	0.000005
GROUND FLOOR	3	DL+LL+RSY Max	-0.018000	0.051000	0.000153	-0.000089	0.000003
GROUND FLOOR	4	DL+LL+RSY Max	-0.006000	0.074000	0.000091	-0.000036	0.000003
GROUND FLOOR	5	DL+LL+RSY Max	-0.004000	0.062000	0.000088	-0.000053	0.000003
GROUND FLOOR	6	DL+LL+RSY Max	-0.004000	0.077000	0.000099	-0.000034	0.000002
GROUND FLOOR	28	DL+LL+RSY Max	-0.020000	0.039000	0.000044	0.000014	0.000001
GROUND FLOOR	19	DL+LL+RSY Max	0.007000	0.023000	0.000211	0.000040	0.000003
GROUND FLOOR	20	DL+LL+RSY Max	0.009000	0.052000	0.001941	-0.000387	0.000003
GROUND FLOOR	21	DL+LL+RSY Max	0.008000	0.023000	0.000355	0.000408	0.000003
GROUND FLOOR	27	DL+LL+RSY Max	0.010000	0.052000	-0.000409	-0.000373	0.000003
GROUND FLOOR	29	DL+LL+RSY Max	0.012000	0.069000	-0.000004	0.000049	0.000003
GROUND FLOOR	64	DL+LL+RSY Max	0.011000	0.058000	0.000046	-0.005396	0.000003
GROUND FLOOR	87	DL+LL+RSY Max	0.002000	0.021000	0.000388	-0.000444	0.000003
GROUND FLOOR	42	DL+LL+RSY Max	-0.017000	0.036000	0.000115	0.000166	0.000003
GROUND FLOOR	63	DL+LL+RSY Max	0.000454	0.021000	0.000146	-0.000048	0.000022
GROUND FLOOR	91	DL+LL+RSY Max	-0.017000	0.034000	-0.000689	-0.000397	0.000003
GROUND FLOOR	96	DL+LL+RSY Max	-0.090000	0.165000	0.000187	-0.000024	0.000008
GROUND FLOOR	152	DL+LL+RSY Max	-0.011000	0.030000	-0.000873	-0.000841	0.000003
GROUND FLOOR	143	DL+LL+RSY Max	-0.019000	0.036000	0.000011	-0.000034	0.000017
GROUND FLOOR	67	DL+LL+RSY Max	0.002000	0.066000	0.000089	-0.000028	0.000003
GROUND FLOOR	218	DL+LL+RSY Max	-0.022000	0.037000	0.000087	-0.000167	0.000003
GROUND FLOOR	219	DL+LL+RSY Max	-0.023000	0.038000	0.000238	-0.000006	0.000003
GROUND FLOOR	220	DL+LL+RSY Max	-0.021000	0.036000	-0.000213	-0.000750	0.000003
GROUND FLOOR	283	DL+LL+RSY Max	0.009000	0.035000	0.000351	0.001093	0.000003
GROUND FLOOR	285	DL+LL+RSY Max	0.010000	0.043000	-0.000083	0.001731	0.000003
GROUND FLOOR	287	DL+LL+RSY Max	-0.022000	0.037000	-0.000002	-0.000263	0.000003
GROUND FLOOR	305	DL+LL+RSY Max	0.011000	0.062000	0.000221	-0.002133	0.000003
GROUND FLOOR	373	DL+LL+RSY Max	0.047000	0.124000	0.000065	0.000008	0.000032
GROUND FLOOR	1024	DL+LL+RSY Max	-0.011000	0.030000	-0.000739	-0.000770	0.000003
GROUND FLOOR	1003	DL+LL+RSY Max	-0.011000	0.029000	-0.000769	-0.000811	0.000003

11.2.6 CONCLUSIONS

In the above we have discussed various problems which we generally encounter in the design of tall buildings trying to satisfy both the interests of the architect and the constraints of codes of practice. Certain modifications in the analysis have been suggested to take care of both, like adding a robust shear wall after reducing the lateral shear carried by the floating columns to zero, a new approach to consider the fundamental torsional component, considering the structure in two parts to improve the mass participation factor and finally introducing a structural joint, in a long narrow building which would serve to consider the building as two shorter buildings.

11.3 PROJECT - 02

11.3.1 INTRODUCTION

The project sets out the basic parameters used for the structural design of a proposed building at Doddanekkundi, Bengaluru. The structural system adopted is as below based on preliminary analysis and design.

The building involves two towers:

The east side tower consisting of 2 basements, ground floor and 9 upper floors and terrace, and the west side tower consisting of 2 basements, ground floor and 9 upper floors and terrace.

Floor-to-floor height for both the towers as per architectural drawings are as follows:

- Lower basement = 3.45 m.
- Upper basement at tower area = 4.65 m.
- Ground and typical floors = 4.0 m.
- Terrace to service (LMR/SHR/ OHT) = 4.5 m.
- The overall height of the structure from basement-2 [RL = 888.4m (FFL)] to terrace [RL = 936.5 m (FFL)] is 48.10 m.

The structural system proposed for this project consists of a flat slab with drop supported on RC column/shear wall.

11.3.2 ANALYSIS

The building is being analyzed to check and balance the center of mass and rigidity. At the initial stage of the analysis, the building configuration was such that the center of mass and the center of rigidity were very far from one another causing eccentricity in the building system which in turn causes the structural instability.

Due to the eccentricity in the building system, the horizontal displacements of a particular floor plate were exceeding the limits and causing torsional irregularities in the building. This has been overcome by altering the structural arrangement of the shear walls and the other structural elements. The brief details of the same are given below. See Figs. 11.12, 11.13 and 11.14.

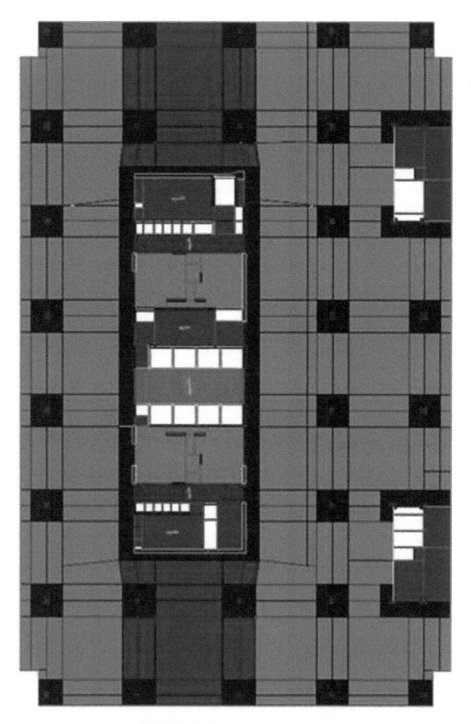

FIGURE 11.12: Analysis model - plan

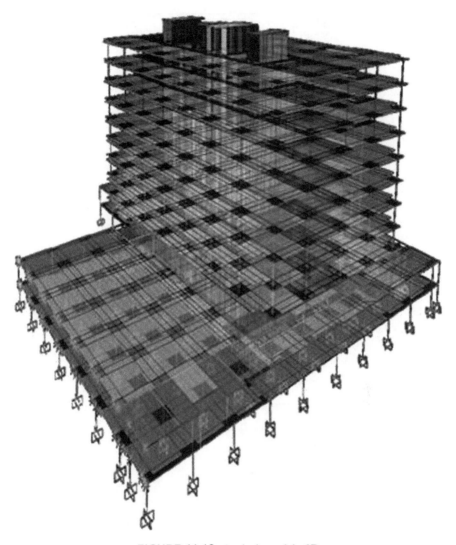

FIGURE 11.13: Analysis model - 3D

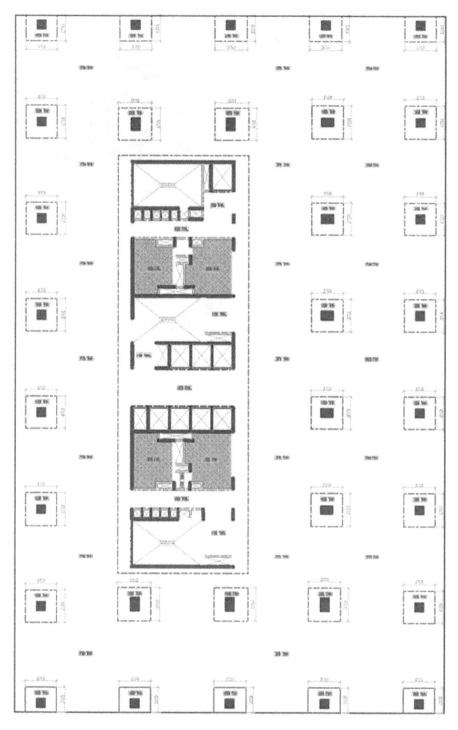

FIGURE 11.14: Structural plan - case 01

Table 14

Case 01 : Centre of Mass and Rigidity							
Storey	Diaphragm	XCR	XCM	Diff.	YCR	YCM	Diff.
		m	m		m	m	
TERRACE	D1	24.37	18.74	5.63	46.21	44.72	1.49
NINTH FLOOR	D1	22.78	16.25	6.53	44.77	44.66	0.11
EIGHTH FLOOR	D1	22.44	16.21	6.23	44.82	44.37	0.45
SEVENTH FLOOR	D1	22.47	16.12	6.35	44.89	44.37	0.52
SIXTH FLOOR	D1	22.83	15.98	6.85	44.96	44.67	0.29
FIFTH FLOOR	D1	22.49	16.18	6.31	45.05	44.37	0.68
FOURTH FLOOR	D1	22.54	16.16	6.38	45.18	44.37	0.81
THIRD FLOOR	D1	22.57	16.04	6.53	45.33	44.4	0.93
SECOND FLOOR	D1	22.56	16.08	6.48	45.5	44.37	1.13
FIRST FLOOR	D1	24.83	18.88	5.95	45.98	44.46	1.52

Table 15

Case 02 : Centre of Mass and Rigidity							
Storey	Diaphragm	XCR	XCM	Diff.	YCR	YCM	Diff.
		m	m	m	m	m	m
TERRACE	D1	22.54	21.56	0.98	44.81	43.98	0.83
NINTH FLOOR	D1	22.70	22.84	-0.14	44.85	44.66	0.19
EIGHTH FLOOR	D1	22.93	22.51	0.42	44.90	44.37	0.53
SEVENTH FLOOR	D1	23.19	22.54	0.66	44.97	44.37	0.59
SIXTH FLOOR	D1	23.46	22.92	0.54	45.03	44.70	0.33
FIFTH FLOOR	D1	23.74	22.56	1.18	45.12	44.37	0.75
FOURTH FLOOR	D1	24.02	22.60	1.42	45.23	44.37	0.86
THIRD FLOOR	D1	24.24	22.60	1.64	45.38	44.37	1.01
SECOND FLOOR	D1	24.40	22.03	2.37	45.58	44.16	1.41
FIRST FLOOR	D1	24.39	25.39	-1.00	45.94	44.14	1.80

The tables above clearly show that the structural system in Case 01 is more eccentric in the X-direction than the structural system in Case 02.

Table 16

| Case 01 : Modal Direction Factors ||||||
Case	Mode	Period	UX	UY	RZ
		sec			
Modal	1	1.579	0.002	0.154	0.844
Modal	2	1.538	0.016	0.832	0.151
Modal	3	1.19	0.983	0.017	0.001
Modal	4	0.475	0.003	0.918	0.08
Modal	5	0.419	0.016	0.103	0.881
Modal	6	0.31	0.225	0.133	0.641
Modal	7	0.299	0.772	0.05	0.178
Modal	8	0.283	0.298	0.023	0.679
Modal	9	0.272	0.036	0.812	0.152
Modal	10	0.209	0.204	0.756	0.041
Modal	11	0.192	0.615	0.297	0.088
Modal	12	0.185	0.259	0.666	0.074
Modal	13	0.178	0.016	0.001	0.983
Modal	14	0.156	0.449	0.372	0.179
Modal	15	0.148	0.001	0.005	0.995
Modal	16	0.134	0.003	0.332	0.665
Modal	17	0.129	0.061	0.469	0.47
Modal	18	0.122	0.024	0.568	0.408
Modal	19	0.115	0.002	0.361	0.637
Modal	20	0.113	0.702	0.002	0.297
Modal	21	0.112	0.098	0.016	0.886
Modal	22	0.107	0.142	0.108	0.75
Modal	23	0.102	0.351	0.043	0.606
Modal	24	0.1	0.068	0.125	0.808
Modal	25	0.098	0.163	0.792	0.045
Modal	26	0.094	0.031	0.3	0.669
Modal	27	0.089	0.138	0.099	0.763
Modal	28	0.088	0.005	0.029	0.967
Modal	29	0.086	0.015	0.133	0.853
Modal	30	0.085	0.026	0.015	0.959

Earthquake Resistant Design as per IS 1893:2016

Table 17

Case 01 : Modal Participating Mass Ratios											
Mode	Period	UX	UY	Sum UX	Sum UY	RX	RY	RZ	Sum RX	Sum RY	Sum RZ
	sec										
1	1.579	0.0004	0.093	0.0004	0.093	0.059	0.0005	0.3059	0.059	0.0005	0.3059
2	1.538	0.0087	0.4733	0.009	0.5664	0.3618	0.0076	0.0415	0.4208	0.0081	0.3474
3	1.19	0.5318	0.0102	0.5409	0.5765	0.0065	0.453	0.0112	0.4273	0.461	0.3587
4	0.475	4.73E-05	0.151	0.5409	0.7276	0.2274	0	0.027	0.6547	0.461	0.3857
5	0.419	0.0062	0.0086	0.5472	0.7361	0.0087	0.0039	0.2464	0.6634	0.465	0.6321
6	0.31	0.0748	0.0199	0.6219	0.7561	0.0225	0.0729	0.1344	0.6859	0.5379	0.7665
7	0.299	0.1895	0.0037	0.8114	0.7597	0.0037	0.1797	0.036	0.6897	0.7176	0.8025
8	0.283	0.0118	0.0016	0.8232	0.7613	0.0016	0.0114	0.0018	0.6913	0.729	0.8042
9	0.272	0.0035	0.1371	0.8267	0.8985	0.1417	0.004	0.0703	0.833	0.733	0.8745
10	0.209	0.0146	0.0005	0.8413	0.899	0.0022	0.0199	0.002	0.8352	0.7529	0.8766
11	0.192	0.0536	0.0115	0.8948	0.9105	0.0162	0.0749	0.001	0.8514	0.8278	0.8776
12	0.185	0.0083	0.001	0.9031	0.9115	0.0016	0.0121	0.0007	0.8529	0.8399	0.8783
13	0.178	0.001	1.12E-05	0.9042	0.9115	2.64E-05	0.0015	0.0007	0.853	0.8414	0.879
14	0.156	0.0223	0.0098	0.9264	0.9213	0.0133	0.0321	0.0039	0.8662	0.8734	0.8829
15	0.148	0.0003	0.0025	0.9267	0.9238	0.0038	0.0005	0.0012	0.87	0.8739	0.8841
16	0.134	0.0002	0.0004	0.9269	0.9242	0.0007	0.0002	0.0009	0.8708	0.8741	0.885
17	0.129	0.0004	8.17E-07	0.9274	0.9242	1.03E-05	0.0006	0.0012	0.8708	0.8747	0.8862
18	0.122	1.27E-05	0.0011	0.9274	0.9253	0.0011	0	0.0002	0.8719	0.8747	0.8864
19	0.115	1.33E-05	0.0015	0.9274	0.9268	0.0022	3.84E-05	0.0003	0.8741	0.8747	0.8867
20	0.113	3.75E-05	2.83E-05	0.9274	0.9268	9.02E-06	0.0009	3.90E-05	0.8741	0.8756	0.8868
21	0.112	0.0004	0.0003	0.9278	0.9271	0.0004	0.001	0.0006	0.8746	0.8766	0.8873
22	0.107	0.0011	0.0006	0.9289	0.9277	0.0013	0.0018	0.0005	0.8759	0.8784	0.8879
23	0.102	0.0087	0.0012	0.9375	0.9289	0.002	0.0147	0.0861	0.8779	0.8931	0.9739
24	0.1	0.0006	0.002	0.9381	0.9309	0.0033	0.0011	0.0006	0.8811	0.8942	0.9745
25	0.098	0.0064	0.0332	0.9445	0.9641	0.0568	0.0108	0.0011	0.938	0.9051	0.9756
26	0.094	0.0024	0.001	0.9469	0.9651	0.0011	0.0041	4.36E-05	0.939	0.9091	0.9757
27	0.089	0.0167	0.0099	0.9636	0.975	0.0175	0.0308	0.0025	0.9565	0.9399	0.9781
28	0.088	0.0014	0.0002	0.965	0.9752	0.0004	0.0015	0.0003	0.9569	0.9414	0.9784
29	0.086	0.0028	0.0001	0.9678	0.9753	0.0003	0.004	0.0001	0.9572	0.9454	0.9786
30	0.085	1.29E-05	4.34E-05	0.9679	0.9753	0.0001	1.08E-05	0.0001	0.9573	0.9454	0.9786

Table 18

| Case 02 : Modal Direction Factors ||||||
Case	Mode	Period	UX	UY	RZ
		sec			
Modal	1	1.6	0.009	0.911	0.08
Modal	2		0.002	0.082	0.916
Modal	3	1.128	0.989	0.01	0
Modal	4	0.477	0.002	0.986	0.012
Modal	5	0.397	0	1	0
Modal	6	0.395	0.001	0.017	0.981
Modal	7	0.294	0.049	0.652	0.299
Modal	8	0.288	1	0	0
Modal	9	0.267	0.971	0.024	0.005
Modal	10	0.264	0	0	1
Modal	11	0.236	0.019	0.762	0.218
Modal	12	0.205	0.158	0.716	0.125
Modal	13	0.198	0.634	0.252	0.114
Modal	14	0.194	0.003	0.017	0.979
Modal	15	0.181	0.025	0.166	0.809
Modal	16	0.163	0.035	0.563	0.402
Modal	17	0.143	0.128	0.052	0.82
Modal	18	0.133	0.001	0.008	0.992
Modal	19	0.127	0.831	0.041	0.128
Modal	20	0.122	0.061	0.754	0.185
Modal	21	0.121	0.663	0.158	0.179
Modal	22	0.115	0.342	0.004	0.654
Modal	23	0.113	0.327	0.019	0.655
Modal	24	0.111	0.194	0.004	0.802
Modal	25	0.106	0.216	0.01	0.775
Modal	26	0.098	0.033	0.548	0.419
Modal	27	0.098	0.023	0.197	0.78
Modal	28	0.093	0	0.042	0.958
Modal	29	0.088	0.001	0.63	0.369
Modal	30	0.086	0.014	0.34	0.647

Earthquake Resistant Design as per IS 1893:2016

Table 19

Trial 2: Modal Participating Mass Ratios

Mode	Period sec	UX	UY	Sum UX	Sum UY	RX	RY	RZ	Sum RX	Sum RY	Sum RZ
1	1.6	0.0055	0.6204	0.0055	0.6204	0.294	0.0034	0.0582	0.294	0.0034	0.0582
2	1.528	0.0008	0.051	0.0063	0.6714	0.0291	0.0008	0.4955	0.3231	0.0042	0.5537
3	1.128	0.5892	0.0073	0.5955	0.6787	0.0031	0.4059	0.0006	0.3262	0.4101	0.5543
4	0.477	0.0004	0.1281	0.5959	0.8068	0.3081	0.0003	0.0005	0.6343	0.4104	0.5548
5	0.397	0	0.0019	0.5959	0.8087	0.0026	0	0.004	0.6369	0.4104	0.5588
6	0.395	0.0003	0.0046	0.5961	0.8133	0.0064	7.48E-06	0.1574	0.6433	0.4104	0.7162
7	0.294	0.0092	0.062	0.6054	0.8753	0.1008	0.0119	0.0701	0.7441	0.4224	0.7863
8	0.288	0.002	0	0.6073	0.8753	0	0.0027	0.0004	0.7441	0.425	0.7867
9	0.267	0.1678	0.0011	0.7751	0.8764	0.0017	0.1827	0.0016	0.7459	0.6077	0.7883
10	0.264	0	0	0.7751	0.8764	0	0	0.0001	0.7459	0.6077	0.7884
11	0.236	0.001	0.0328	0.7761	0.9092	0.0476	0.0022	0.0441	0.7935	0.6099	0.8325
12	0.205	0.0045	0.0029	0.7806	0.9121	0.0059	0.0078	0.002	0.7994	0.6176	0.8345
13	0.198	0.0342	0.0002	0.8148	0.9124	0.0008	0.0561	0.0001	0.8002	0.6737	0.8345
14	0.194	0.0008	0.0006	0.8156	0.913	0.0016	0.0013	0.0022	0.8018	0.675	0.8367
15	0.181	0.0039	0.0028	0.8195	0.9158	0.0066	0.0061	0.0283	0.8084	0.6811	0.8651
16	0.163	0.0047	0.0264	0.8242	0.9421	0.0556	0.0074	0.0202	0.864	0.6884	0.8852
17	0.143	0.0138	0.0014	0.838	0.9435	0.0029	0.0227	0.0108	0.8669	0.7111	0.8961
18	0.133	0.0002	3.77E-05	0.8382	0.9435	0.0001	0.0003	0.0008	0.867	0.7114	0.8968
19	0.127	0.0212	2.38E-05	0.8594	0.9435	0.0001	0.0306	0.0016	0.8671	0.742	0.8984
20	0.122	0.003	0.0009	0.8624	0.9445	0.0016	0.0042	0.0034	0.8687	0.7462	0.9018
21	0.121	0.0107	0.0002	0.8732	0.9447	0.0003	0.0136	0.0004	0.869	0.7597	0.9022
22	0.115	0.0019	3.03E-05	0.8751	0.9447	0.0001	0.0028	0.0021	0.8691	0.7625	0.9043
23	0.113	0.0001	0.0003	0.8752	0.945	0.0006	6.02E-06	0.0015	0.8697	0.7625	0.9058
24	0.111	0.0069	3.28E-05	0.882	0.9451	0.0001	0.012	0.0003	0.8698	0.7746	0.9061
25	0.106	0.0008	0.0001	0.8828	0.9451	0.0001	0.0021	0.0001	0.8699	0.7766	0.9062
26	0.098	0.0002	0.0005	0.883	0.9456	0.0007	0.0006	0.001	0.8706	0.7772	0.9072
27	0.098	0.0002	0.0019	0.8832	0.9475	0.0039	0.0008	0.0023	0.8745	0.778	0.9095
28	0.093	8.97E-07	2.55E-05	0.8832	0.9475	0.0001	8.61E-07	0	0.8746	0.778	0.9095
29	0.088	3.15E-06	0.0001	0.8832	0.9476	0.0001	1.43E-06	2.58E-05	0.8746	0.778	0.9096
30	0.086	0.0001	0.0111	0.8833	0.9587	0.0268	0.0001	0.0262	0.9014	0.7781	0.9358

11.3.3 CONCLUSIONS

Observing the tables above the following conclusions can be drawn. In Case 01 the results show that the center of mass and center of rigidity are very far from one another, and also torsion is dominating in the first mode. After altering the core wall of the structure, the results for Case 02 show that the center of rigidity and center of mass are brought closer, and torsion has been shifted to the second mode.

11.4 PROBLEMS

Q. 01. Find out the natural frequency for the first 10 modes for the symmetric building configuration given. For the analysis, consider the parameters given below:

Number of floors and heights of each floor: 3 basements (4.0 m), ground floor (5 m), 14 typical floors (4.5 m each). See fig. 11.15.

Consider the thickness of the shear walls 300 mm, if not given.

Floor Loads - Live load - 4 kN/m^2, Finishes - 1.5 kN/m^2, Services - 0.5 kN/m^2.

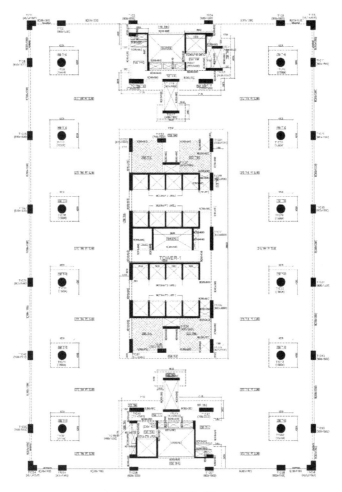

FIGURE 11.15: Plan

Q. 02. For the above problem, find out the earthquake response for the first 10 modes considering earthquake zone II.

Q. 03. Find out the natural frequency for the first 10 modes for the eccentric building configuration given. For the analysis, consider the parameters given below:

Number of floors and heights of each floor - 2 basements (3.6 m), ground floor (4.5 m), 11 typical floors (3.9 m each). See Plan of the building. Fig. 11.16.

Consider the thickness of the shear walls is 300 mm, if not given.

Floor Loads - Live load - 4 kN/^{2}, Finishes - 1.5 kN/^{2}, Services - 0.5 kN/^{2}.

Earthquake Resistant Design as per IS 1893:2016

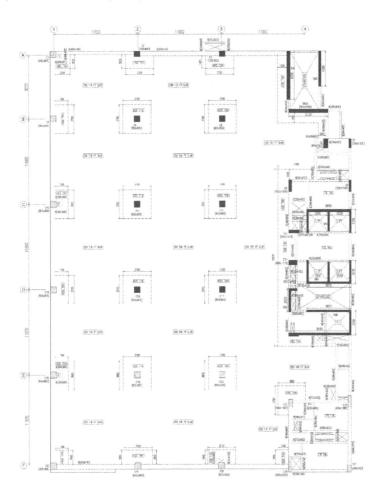

FIGURE 11.16: Plan

Q. 04. For the above problem, find out the earthquake response for the first 10 modes considering earthquake zone II.

Q. 05. Find out the natural frequency and mass participation factors for the first 10 modes for the given building configuration with podium. For the analysis, consider the parameters given below. See plan of the building. Fig. 11.17.

Number of floors and heights of each floor - 2 basements (3.6 m), ground floor (5 m), 8 typical floors (4.5 m each).

Column and beam sizes - tower columns (1000×1200 mm), podium columns (800×800 mm), beam sizes (450×650 mm), thickness of the shear walls (300 mm).

Floor Loads - Live load - 4 kN/^{2}, Finishes - 1.5 kN/^{2}, Services - 0.5 kN/^{2}.

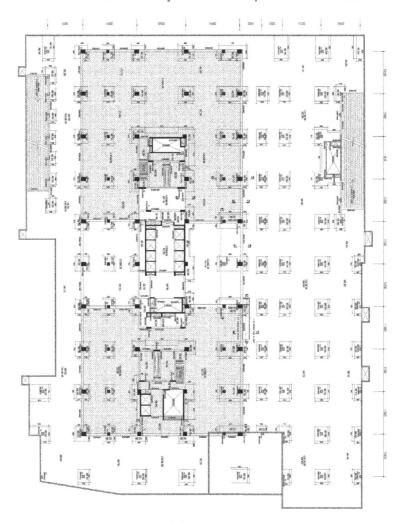

FIGURE 11.17: Plan of the building

Q. 06. For the above problem, find out the natural frequency and mass participation factors for the first 10 modes by assigning the fixed support at the podium level.

Q. 07. Find out the natural frequency and mass participation factors for the first 10 modes for the given building configuration with 2 towers and podium. For the analysis, consider the parameters given below.

Number of floors and heights of each floor - 2 basements (3.6 m), ground floor (4 m), 10 typical floors (3.75 m each). See the plan. Fig. 11.18. for the tower and Fig. 11.19 for the podium.

Floor Loads - Live load - 4 kN/l^2, Finishes - 1.5 kN/l^2, Services - 0.5 kN/l^2.

Earthquake Resistant Design as per IS 1893:2016

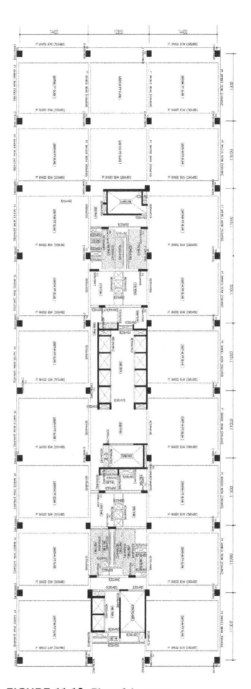

FIGURE 11.18: Plan of the tower

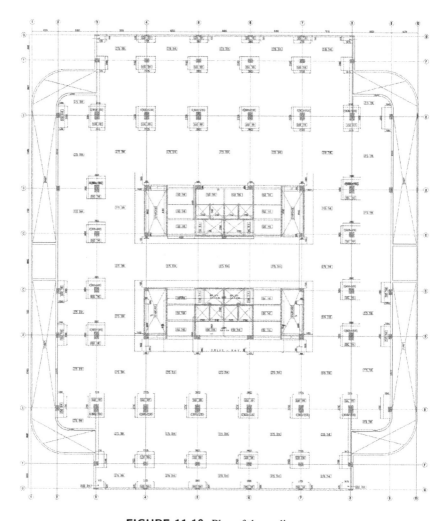

FIGURE 11.19: Plan of the podium

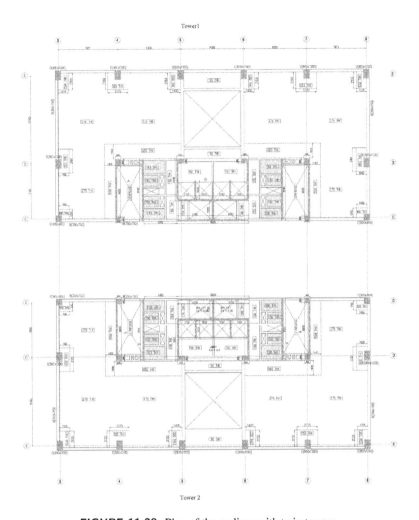

FIGURE 11.20: Plan of the podium with twin towers

Q. 08. For the above problem, find out the natural frequency and mass participation factors for the first 10 modes by providing an expansion joint between the two towers. See Fig. 11.20.

Q. 09. Find out the torsional irregularities for the given narrow building configuration. For the analysis, consider the parameters given below.

Number of floors and heights of each floor - 1 basements (3.6 m), ground floor (3 m), 12 typical floors (3 m each). Thickness of the shear walls is 300 mm. See the Fig. 11.21.

Floor Loads - Live load - 2 kN/^{2}, Finishes - 1.5 kN/^{2}.

FIGURE 11.21: Plan of the narrow building

Q. 10. In the above, if torsional irregularity exists, find out the ways to rectify it and justify the same.

12 Miscellaneous Aspects

12.1 INTRODUCTION

In the last seven years, the author have been guiding the dissertation projects of graduate students in the Department of Civil Engineering in the Oxford College of Engineering, Bangalore. Recently, the author happened to suggest a few interesting problems concerning the current interest in earthquake resistant design. These problems relate to torsion response in buildings asymmetric in plan, rehabilitation of damaged reinforced concrete (RCC) buildings, hybrid reinforced concrete and steel buildings, and using Python to obtain response spectrum. The intention of using Python is that it is an open source language which can be freely downloaded. Python circumvents use of expensive commercial software. Python is used for the first time and the author's hope is that there will be more research in Python. The results from the various interesting disssertation projects done various post graduate students under the guidance of the author along with Prof. Amarnath at Oxford college of engineering.

12.2 RETROFITTING METHODS IN REINFORCED CONCRETE STRUCTURES

The following work is a part of the dissertation of student, Gayathri, R. Retrofitting is the process of strengthening structural elements of the existing damaged building. The strength of the whole structure can be improved by introducing new structural elements or strengthening the existing structural elements.

The retrofitting becomes necessary to improve the performance of structures, which lose strength due to deterioration or which have exceeded their expected life. The success of retrofitting depends on the actual cause and the measures adopted to prevent its further deterioration. To improve the strength of existing structures, the following are the two ways:

(i) Structure-level retrofit.
(ii) Member-level retrofit.

12.2.1 STRUCTURE-LEVEL RETROFIT

In this way of retrofitting, the strength of the structure is enhanced by introducing new structural members to an existing damaged building. Some of the structure-level retrofitting methods are described below.

12.2.2 ADDITION OF SHEAR WALLS

Shear walls in a framed structure are used to resist earthquake forces. Shear walls can also be added for retrofitting.

12.2.3 BASE ISOLATORS

Base isolators are introduced between the base and the main structure so as to reduce the base shear.

12.2.4 ADDITION OF STEEL BRACING

A braced frame is a structural member that will be added to the existing building to reduce the effect of earthquake and wind forces. The addition of steel bracing is an efficient method of retrofitting.

12.2.5 MEMBER-LEVEL RETROFIT

Here, structural elements are retrofitted. New structural steel sections are used to connect the two parts of the damaged member after removing the damaged portion. The damaged member can be either reinforced concrete or structural steel. If the damaged member is structural steel, it is easier to use structural steel as a connecting member, because steel-to-steel connections are well established. If the damaged member is reinforced concrete then the connection between steel and reinforced concrete is quite complex, in which case concrete filled steel tubes can be used as connecting members. The advantage here is the concrete filled steel tube (CFST) with reinforcement inside will help connections either with reinforced concrete or structural steel. If the damaged member is structural steel, the steel tube surrounding the CFST will help in connecting the steel tube to the structural steel member. The reinforcement in the concrete can also be used in addition. If the damaged member is reinforced concrete, then the concrete inside the steel tube with reinforcement can be connected to the reinforced concrete. Here in the following sections, both methods are described.

SECTION ENLARGEMENT OR JACKETING

This is the most common method of retrofitting beams and columns of an RCC structure. Additional reinforcement is provided around the beam and column section like a cage (Fig. 12.1). So, it will increase the size of the beam and column and will help to increase the strength of the structural member. Here the existing concrete and the enlargement should be bonded nicely to produce a monolithic member. This method is more economical.

Example of G+9 RC building retrofitted by two methods. Safe bearing capacity (SBC) of soil = 200 KN/mm^2. Total height of the building = 31.5 m. An already existing damaged G+9 building of plan dimension 15 m × 15 m is considered. It is situated in zone IV. The soil condition is medium. The overall height of the building

Miscellaneous Aspects

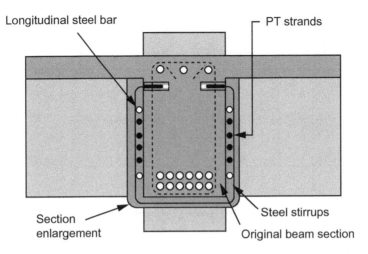

FIGURE 12.1: Section enlargement of beam

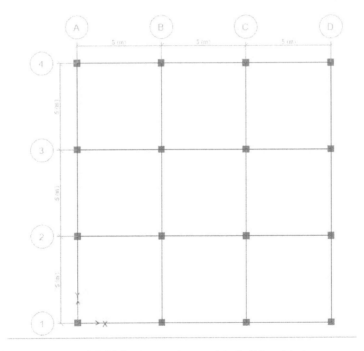

FIGURE 12.2: Column layout of the building (plan)

is 31.5 m with basement height of 4.5 m and the typical, other floor height of 3 m. The building consists of columns of size 400 mm × 400 mm and beams of size 230 mm × 300 mm, respectively. The wall thickness and slab thickness of the building are 200 mm and 125 mm, respectively. The plan of the layout of the columns is as shown in Fig. 12.2 and the corresponding elevation in Fig. 12.3.

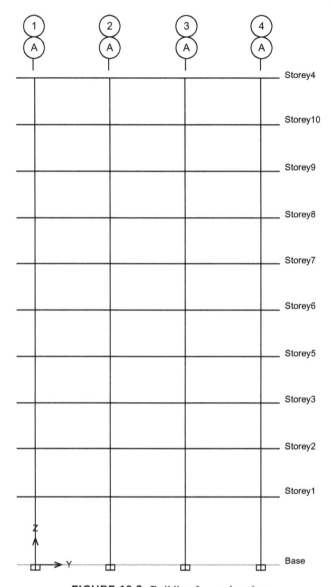

FIGURE 12.3: Building frame elevation

Some columns and beams at the basement level are damaged in the above building. So, the damaged beams and columns at the basement level are retrofitted. The damaged portions of the beam and column are completely replaced by two systems:

(i) Replacing them by steel I sections.
(ii) Replacing them by concrete filled steel tube sections.

Miscellaneous Aspects

Then the building is analyzed with its retrofitted elements in ETABS 2016 software with response reduction factor 5. Both static and dynamic analysis and results are compared.

Design Parameters:
Details of an RCC framed structure.

Concrete mix for slabs: M 25
Concrete mix for beams: M 30
Concrete mix for columns: M 35

Loading Parameters:
(a) Dead Load:

The dead load of the structural member will be considered on the basis of the following properties:

Density of plain concrete: 24.0 kN/m^3
Density of reinforced concrete: 25.0 kN/m^3
Density of steel: 78.5 kN/m^3

(b) Live Load:

The following are the imposed loads considered in addition to the dead loads of the structure:

Live load on basement and typical floor slab: 3.2 kN/m^2
Live load on terrace slab: 1.5 kN/m^2
Floor finish on typical slab: 1.5 kN/m^2
Floor finish on terrace slab: 0.8 kN/m^2

(c) Wall Load over Beams:

Wall thickness = 200 mm
Masonry wall height = 3 m
Self-weight of masonry = (0.20*21.5*3) = 12.9 kN/m
Plastering = (0.04*20*3) = 2.4 kN/m
Total load = 15.3 kN/m

(d) Earthquake Consideration:

Type of Soil: Medium
Importance Factor: 1
Response Reduction Factor: 5
Zone: IV

Two Methods of Retrofitting

The retrofitting is important to enhance the strength and performance of the structure. Retrofitting has to be done for the structures that exceeded their expected life and lost their strength due to deterioration. The success of retrofitting depends on the method of retrofitting chosen. Here for this damaged building, steel sections are used for retrofitting because of the following reasons:

(i) The steel sections have high strength.
(ii) The steel sections can be easily handled and transported.
(iii) They are more ductile.
(iv) They can dissipate considerable energy under cyclic loading.

The existing damaged building is retrofitted by two systems:

(a) Retrofitted with steel I sections.
(b) Retrofitted with concrete filled steel tube sections.

In this building, the beams and columns at the basement level are damaged. So the steel sections are more comfortable for retrofitting. The methodology of using two different steel sections for retrofitting is described below.

Retrofitted with Steel I Section:
Steel section is used for connection.

Beam Retrofitting

1. The size of steel I section - 230 × 300 mm.
2. Total depth - 300 mm.
3. Total width - 230 mm.
4. Top flange thickness - 25 mm.
5. Web thickness - 13 mm.
6. Bottom flange thickness - 25 mm.
7. Fillet radius - 0 mm.

Column Retrofitting

1. The size of steel I section - 400 × 400 mm.
2. Total depth - 400 mm.
3. Total width - 400 mm.
4. Top flange thickness - 25 mm.
5. Web thickness - 13 mm.
6. Bottom flange thickness - 25 mm.
7. Fillet radius - 0 mm.

Procedure followed for retrofitting is as follows:

Step 1:

The damaged beam in a frame is shown in Fig. 12.4. In this type of retrofitting, the damaged portions of the beam and column are completely replaced by steel I section. The middle portion of the beam is damaged completely. The middle one-third portion of the beam is completely removed.

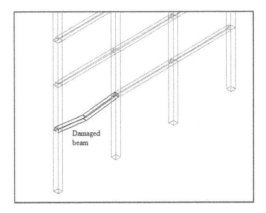

FIGURE 12.4: Damaged beam

Step 2:

The remaining RCC portions are plastered well.

Step 3:

The steel I section and the existing RCC beam section are connected by end plates and anchor steel stud rods. The steel stud rods of diameter 16 mm with a development length of 0.9 m are drilled into the existing RCC beam sections on both sides (Fig. 12.5).

Step 4:

The endplates of thickness 35 mm are welded to the steel stud rods. The drilled holes are filled with epoxy grout.

Step 5:

The steel I section of 230 × 300 mm is welded to the end plates.

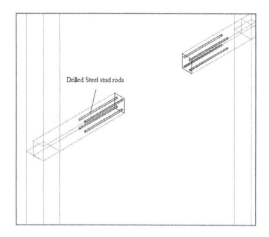

FIGURE 12.5: Drilled steel stud rods

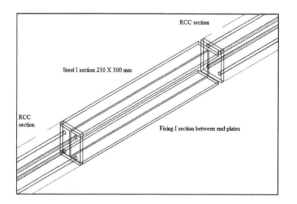

FIGURE 12.6: Steel I section fixed between RCC section

Step 6:

The retrofitted beam with Steel I section is as shown in Figure 12.6.

In column retrofitting, the procedure followed is the same as that of beam retrofitting. Here steel I section and the existing RCC column sections are connected by using base plates and anchor steel stud rods as shown in Fig. 12.7. The dimensions of steel section used in column retrofitting are shown in Fig. 12.7. The steel stud rods are drilled into the existing RCC column sections. The base plates of thickness 30 mm are welded to the steel stud rods. The drilled holes are filled with epoxy grout. The ends of steel I section are welded to the base plates.

Miscellaneous Aspects

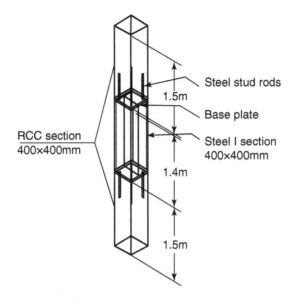

FIGURE 12.7: Column retrofitted with steel I section

Retrofitted with Concrete Filled Steel Tube Section:

CFSTs are composite members having both steel and concrete material. The CFST consists of steel outside with concrete inside. It enhances the strength and stiffness of the section. The steel located on the outer surface can bear the stress due to tension. The modulus of elasticity of steel is much more than that of concrete so that the stiffness of CFST is greatly enhanced. The concrete filled in the middle can withstand the stress due to compression. The buckling of steel is prevented and it increases the compressive strength. Here the beams and columns are retrofitted in the same procedure as that of retrofitting with steel I section as described above.

Section properties of concrete filled steel tube section are as follows:

Beam Retrofitting

1. The size of CFST section - 230 × 300 mm.
2. Total depth - 230 mm.
3. Total width - 300 mm.
4. Top flange thickness - 13 mm.
5. Web thickness - 13 mm.

Column Retrofitting

1. The size of CFST section - 400 × 400 mm.
2. Total depth - 400 mm.

3. Total width - 400 mm.
4. Top flange thickness - 13 mm.
5. Web thickness - 13 mm.

Material Properties

1. Grade of concrete in CFST section M25.
2. Grade of Steel in CFST section Fe250.

The retrofitted beam and column with CFST section. The procedure of retrofitting is the same as that adopted for steel I section.

Free Vibration Analysis of Existing Building:
The existing G+9 building is analyzed by free vibration analysis. After analysis, the natural frequencies and mode shapes of the structure are given below, in Table 12.1.

(a) Natural Frequencies

Table 12.1 Period and natural frequency

Mode	Period(sec)	Natural Frequency(cyc/sec)
1	1.897	0.527
2	1.897	0.527
3	1.617	0.618
4	0.608	1.644
5	0.608	1.644
6	0.519	1.927
7	0.339	2.949
8	0.339	2.949
9	0.291	3.439
10	0.222	4.5
11	0.222	4.5
12	0.191	5.247

The following models are studied:

Model 1- Existing building before retrofitting.
Model 2- Building retrofitted with steel I section.
Model 3- Building retrofitted with steel tube filled with concrete.

Miscellaneous Aspects

The above three models are analyzed using both the static and dynamic methods using ETABS 2016. Results are shown in Tables 12.2 through 12.5.

(b) Base Shear

Table 12.2 Base shear (kN)

Type of Building	Equivalent Static Analysis		Response spectrum analysis	
	EQX	EQY	RSX	RSY
Model 1	-356.22	-356.22	713.41	713.41
Model 2	-354.54	-354.54	716.24	716.23
Model 3	-355.46	-355.46	723.49	723.49

(c) Storey Drift

Table 12.3 Storey drift

Type of Building	Equivalent Static Analysis		Response spectrum analysis	
	EQX	EQY	RSX	RSY
Model 1	0.002183	0.002183	0.003935	0.003935
Model 2	0.002169	0.002169	0.003978	0.003978
Model 3	0.002158	0.002158	0.003874	0.003874

(d) Max Storey Displacement

Table 12.4 Max storey displacement (mm)

Type of Building	Equivalent Static Analysis		Response Spectrum analysis	
	EQX	EQY	RSX	RSY
Model 1	52.94	52.94	82.76	82.76
Model 2	52.25	52.25	83.32	83.32
Model 3	52.13	52.13	83.07	83.07

(e) Comparison of Deflection, Shear Force, Moment at Joint

Table 12.5 Moment, shear and deflection at joint

Type of Building	Moment at Joint kN-m	Shear Force at joint kN	Deflection at joint mm
Model 1	24.57	26.29	1.3
Model 2	10.61	19.29	0.4
Model 3	10.11	18.73	0.4

The deflection, moment and shear forces at the joint between RCC and steel section for the above three models are tabulated and compared.

Here the deflection at the joint of a sample damaged beam, Model 1 (without retrofitting) is 1.3 mm, and the deflection is reduced to 0.4 mm in both Model 2 and Model 3.

12.3 RESPONSE SPECTRUM ANALYSIS USING PYTHON

12.3.1 ABOUT THE PROGRAMMING LANGUAGE: PYTHON

The following work is a part of the dissertation of student Swathi Ricke. Python is an elucidated high-level programming language for general-purpose programming in all fields of engineering. It was coined and developed by Guido van Rossum in 1991. Python has a design philosophy that allows code readability, and a syntax allowing programmers to design and develop concepts in fewer lines of code, notably using significant whitespace. Python highlights a dynamic type system and automatic memory management. It supports multiple programming paradigms with object-oriented, imperative, functional, and procedural features, and facilitates a huge as well as comprehensive standard library. Python interpreters are available for various operating systems. Python is organized by the non-profit Python Software Foundation. Like many fields of engineering, the various branches of civil engineering also incorporate data science applications; therefore Python becomes the most approved programming language in data science.

12.3.2 HISTORY OF PYTHON

Python implementation commenced in December 1989 by Guido van Rossum. Python was influenced by other programming languages like C, C++, Java, Perl, and Lisp. Its accessible and user-friendly syntax made Python more popular. Python's big benefit is that it is simple to learn and is a smaller language than numerous other software programming languages like Java, C or C++ programs. Python is an active software programming language that is appreciable for the programming of script which is helpful for setting up automated procedures in civil engineering projects.

12.3.3 APPLICATION OF PYTHON IN CIVIL ENGINEERING

Like many fields of engineering, civil engineering also involves data science applications. Python is the most popular programming language in data science.

The applications of data science in civil engineering are as follows:

• Population forecasting for urban planning, water supply and sewerage systems.

• Risk assessment and mitigation such as prediction of floods, earthquakes, and cyclones.

• Structural health monitoring.

• Prediction of traffic trends in highway engineering.

• Soil simulation and modelling in geotechnical engineering.

• Finite element method (FEM) applications in structural engineering.

• Construction management.

• Machine learning (ML) applications include automation in structural design and drawings.

Python has also made its application in civil engineering for automation tasks like calculating bending moment, shear force, reactions at supports, and so on. One can use any code as a module to refer to while designing complex problems like analysis and design of multi-storey buildings using Python. Here it has been used almost for the first time to do response spectrum analysis of a multi-storey building. The world is changing fast and dynamically due to many factors. Artificial intelligence machine learning dominates all branches of engineering and more in civil engineering.

We should be ready for the next big challenge: automation in the civil engineering industry.

12.3.4 PYTHON ARCHITECTURE

Syntax and semantics:

Python is a simply readable language. Its format is visually unlittered, and it often uses English keywords wherever alternative languages use punctuation. Unlike several alternative languages, it does not use curly brackets to delimit blocks, and semicolons which are used after statements are optional. It has a little amount of syntactic exceptions and special cases than C or Pascal.

1. Indentation:

Python uses white-space indentation, instead of curly brackets or keywords, to delimit blocks.

2. Statements and control flow:

- The statement 'if' conditionally executes a set of codes, along with 'else' and 'elif' (a contraction of else-if).

- The 'for' statement iterates over an iterable object.

- The 'while' statement executes a set of codes as long as its condition is true.

- The 'class' statement executes a set of codes and attaches its local namespace to a class, for utilizing in object-oriented programming.

- The 'def' statement defines a function or method.

- The 'yield' statement returns a value from a generator function.

- The 'import' statement is used to import modules whose functions or variables can be used in the current program.

- The 'print' statement was replaced to the print() function in Python 3.0.

3. Expressions:

- Some Python expressions are identical to languages such as C and Java, while some are not.

- Addition, subtraction, and multiplication are similar, but the behavior of division varies. There are two types of divisions in Python, i.e., integer division and floor division. Python has also added the ** operator for exponentiation.

- It has been intended to be used by libraries such as NumPy for matrix multiplication.

- In Python, == correlates by value, versus Java, which compares numerics by value and objects by reference.

- It uses the words, and, or, not, for its Boolean operators rather than the symbolic, ||, ! used in Java and C.

- Conditional expressions are written as x if c else y.

- Python makes a difference between lists and tuples. Lists are composed as [1, 2, 3] are mutable and cannot be used as the keys of dictionaries as dictionary keys must be immutable in Python.

- Tuples are drafted as (1, 2, 3) are immutable and so are often used as the keys of dictionary.

- It has a "string format" operator %.

4. Typing:

Python allows programmers to define their own required types using classes, which are most often used for object-oriented programming.

5. Mathematics:

Python has similar C language arithmetic operators (+, -, *, /, %). It also has ** for exponentiation.

12.3.5 PYTHON LIBRARIES

Python's massive customary library, commonly cited as one of its greatest strengths, provides tools suited to several tasks. Some of the libraries are listed below:

- NumPy- It supports some advance math functionalities to Python.

- SciPy- It is a library consisting of algorithms and mathematical tools for Python.

- matplotlib- It is a numerical plotting library. It is very useful for any data analyzer or any data scientist.

- Pygame- This library helps to achieve your goal of 2D game development.

- pywin32- A Python library providing some useful methods and classes for interacting with windows.

- SymPy- A library that can do algebraic evaluation, differentiation, expansion, complex numbers, etc.

240   Structural Dynamics in Earthquake and Blast Resistant Design

- IPython- It provides completion, history, shell capabilities, and a lot more.

- Pandas- It offers data structures and operations for manipulating numerical tables and time series.

12.3.6 STATIC LOADING PROBLEM USING PYTHON

Problem 1:

A fixed beam of span AB 8 m carrying UDL of 30 kN/m, span BC 12 m carrying point load at a distance of 4 m from B, span CD 6 m carrying UDL of 20 kN/m (Fig. 12.8).

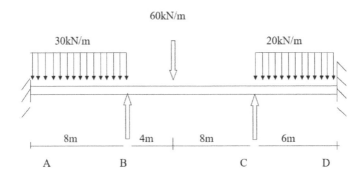

FIGURE 12.8: Fixed beam

Python program:
```
# Fixed beam:
Wab =int(input("span AB Wab: "))
Wbc =int(input("span B Wbc: "))
Wcd =int(input("span CD Wcd: "))
Lab =int(input("Length of span AB Lab: "))
LbcA = int(input("Length in span BC Lbc-A: "))
LbcB = int(input("Length in span BC Lbc-B: "))
Lcd = int(input("Length in span CD Lcd: "))
Iab = float(input("Iab: "))
Ibc = float(input("Ibc: "))
Icd = float(input("Icd: "))

#Free End Moments:
Mfab= -(Wab*Lab**2)/12
Mfba= (Wab*Lab**2)/12
Mfbc= -(Wbc*LbcA*LbcB**2)/((LbcA+LbcB)**2)
```

Mfcb= (Wbc*LbcB*LbcA**2)/((LbcA+LbcB)**2)
Mfcd= -(Wcd*Lcd**2)/12
Mfdc= (Wcd*Lcd**2)/12
print ("Mfab=",Mfab)
print ("Mfba=",Mfba)
print ("Mfbc=",Mfbc)
print ("Mfcb=",Mfcb)
print ("Mfcd=",Mfcd)
print ("Mfdc=",Mfdc)

#Stiffness factor (K):
EI="EI"
Kab = (4*Iab/Lab)
Kbc = (4*Ibc/(LbcA+LbcB))
Kcd = (4*Icd/(Lcd))
print (Kab,Kbc,Kcd)

#Distribution factor(D):
Uba = -(1/2)*(Kab/(Kab+Kbc))
Ubc = -(1/2)*(Kbc/(Kab+Kbc))
Ucb = -(1/2)*(Kbc/(Kcd+Kbc))
Ucd = -(1/2)*(Kcd/(Kcd+Kbc))
print (Uba,Ubc,Ucb,Ucd)
mab=0
mba= (-Mfab+Mfbc)*Uba
mbc= (-Mfab+Mfbc)*Ubc
mcb= (+Mfcb+Mfcd+mbc)*Ucb
mcd= (Mfcb+Mfcd+mbc)*Ucd
mdc=0
print (mba,mbc,mcb,mcd)

#Final moments
Mab= Mfab+(2*mab)+mba
Mba= Mfba+(2*mba)+mab
Mbc= Mfbc+(2*mbc)+mcb
Mcb= Mfcb+(2*mcb)+mbc
Mcd= Mfcd+(2*mcd)+mdc
Mdc= Mfdc+(2*mdc)+mcd
print ("Mab=",Mab,"kNm")
print ("Mba=",Mba,"kNm")
print ("Mbc=",Mbc,"kNm")
print ("Mcb=",Mcb,"kNm")
print ("Mcd=",Mcd,"kNm")
print ("Mdc=",Mdc,"kNm")

Solution:
Final moments
Mab= -112.0 kNm
Mba= 256.0 kNm
Mbc= -289.0 kNm
Mcb= 126.0 kNm
Mcd= -126.0 kNm
Mdc= 27.0 kNm

Problem 2:

A cantilever beam of span AB 4 m carrying a moment 24 kN/m at a distance of 1.5 m from A, span BC 4 m carrying a moment 32 kN/m at a distance of 2 m from C, span CD of 3 m carrying a point load 36 kN at a distance of 1 m from C (Fig. 12.9).

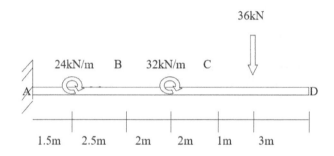

FIGURE 12.9: Cantilever beam

Python Program:
```
#Cantilever beam:
Wab =int(input("Wab: "))
Wbc =int(input("Wbc: "))
Wcd = float(input("Wcd: "))
Lab =int(input("Length of span AB Lab: "))
Lbc = int(input("Length in span BC Lbc: "))
Lcd = float(input("Length in span CD Lcd: "))
Iab = float(input("Iab: "))
Ibc = float(input("Ibc: "))
Icd = float(input("Icd: "))

#Free End Moment:
Mfab= -(Wab*Lab**2)/12
Mfba= (Wab*Lab**2)/12
Mfbc= -(Wbc*Lbc**2)/12
Mfcb= (Wbc*Lbc**2)/12
Mfcd= -(Wcd*Lcd)
```

Mfdc=0
print ("Mfab=",Mfab)
print ("Mfba=",Mfba)
print ("Mfbc=",Mfbc)
print ("Mfcb=",Mfcb)
print ("Mfcd=",Mfcd)
print ("Mfdc=",Mfdc)

#Stiffness factor (K):
EI="EI"
Kab=(4*Iab/Lab)
Kbc = (4*Ibc/(Lbc))
Kcd = (4*Icd/(Lcd))
print (Kab,Kbc,Kcd)

#Distribution factor(D):
Uba = -(1/2)*(Kab/(Kab+Kbc))
Ubc = -(1/2)*(Kbc/(Kab+Kbc))
Ucb = -(1/2)*(Kbc/(Kcd+Kbc))
Ucd = -(1/2)*(Kcd/(Kcd+Kbc))
print (Uba,Ubc,Ucb,Ucd)
mab=0
mba= (-Mfab+Mfbc)*Uba
mbc= (-Mfab+Mfbc)*Ubc
mcb= (+Mfcb+Mfcd+mbc)*Ucb
mcd= (Mfcb+Mfcd+mbc)*Ucd
mdc=0

#Final moments:
Mab= Mfab+(2*mab)+mba
Mba= Mfba+(2*mba)+mab
Mbc= Mfbc+(2*mbc)+mcb
Mcb= Mfcb+(2*mcb)+mbc
Mcd= Mfcd+(2*mcd)+mdc
Mdc= Mfdc+(2*mdc)+mcd
print ("Mab=",Mab,"kNm")
print ("Mba=",Mba,"kNm")
print ("Mbc=",Mbc,"kNm")
print ("Mcb=",Mcb,"kNm")
print ("Mcd=",Mcd,"kNm")
print ("Mdc=",Mdc,"kNm")

Solution:
Final moments:
Mab= -7.152777777777779 kNm
Mba= 5.694444444444445 kNm

Mbc= -5.138888888888889 kNm
Mcb= 3.8888888888888884 kNm
Mcd= -3.888888888888889 kNm
Mdc= 0.5555555555555556 kNm

12.3.7 FREE VIBRATION PROBLEMS

Problem 3:

The steel frame is fixed at the base and has a rigid top that weighs 4.48 kN. Experimentally it has been found that its natural period in lateral vibration is equal to 1/10 of a second. It is required to shorten or lengthen its period by 20% by adding weight or strengthening the columns. Determine needed additional weight or additional stiffness (neglect the weight of the column).

Python program:
```
#w=(2*3.142/T)=(k*g/W)**(1/2)
# T in seconds:
T=0.1
# g in m/sec2:
g=9.81
# Weight in kN:
W=4.48
#To calculate Stiffness:
k= ((2*3.142/T)**2)*W/g
print("k=",k,"kN/mm")
#lengthen the period by 20% by adding weight DW:
Tl=1.2*0.1
DW= ((k*g)/((2*3.142/Tl)**2))-4.48
print("DW=",DW,"lb")
#Shortening the period by 20% by strenghtening the columns by Dk:
Ts=0.8*0.1
Dk= (((2*3.142/Ts)**2)*4.48/g)-k
print ("Dk=",Dk,"lb/in")
```

Solution:
k= 1.791 kN/mm
DW= 1.957 kN
Dk= 1.007 kN/mm

Problem 4:

A vibrating system consisting of a weight of 44.48 N and a spring with stiffness 3.502 N/mm is viscously damped so that the ratio of two consecutive amplitudes is 1.0 to 0.85. Determine the damped and undamped natural frequency, damping ratio and coefficient.

Miscellaneous Aspects

Python program:
```
#weight on the vibrating system in N:
W=44.48
# Stiffness in N/mm:
k=3.502
#two consecutive amplitudes:
a1=1
a2=0.85
#the undamped natural frequency f in cps:
f= (1/(2*3.142))*((k*9.81/W)**(1/2))
print ("f=",f,"cps")
# The logarithmic decrement:
import math
Delta= math.log(a1/a2)
print("Logarithmic decrement=",Delta)
# The damping ratio Dr is given by:
Dr= Delta/(2*3.142)
print("Damping ratio=",Dr)
#damping coefficient C :
C= 2*Dr*((k*W/g)**(1/2))
print("Damping coefficient C=",C,"Ns/mm")
#The natural frequency of damped frequency in rad/sec:
wD= (f*2*3.142)*((1-Dr**2)**(1/2))
print ("Damped frequency wD=",wD,"rad/sec")
```

Solution:
Natural frequency,f= 0.18 cps
Logarithmic decrement= 0.16
Damping ratio= 0.025
Damping coefficient C= 0.199 Ns/mm
Damped frequency wD= 1.13 rad/sec

12.3.8 RESPONSE SPECTRUM ANALYSIS OF BUILDING USING PYTHON

Problem 5: (Figs. 12.10 and 12.11).
The following data are used in the analysis:

- **Section Properties:**
    - Size of beam: 230×450 mm.
    - Size of column: 230×450 mm.
    - Depth of Slab: 150 mm.
    - Thickness of wall: 150 mm.
- **Storey Properties:**
    - Height between the floors: 3.5 m.
    - Number of storey: 1.

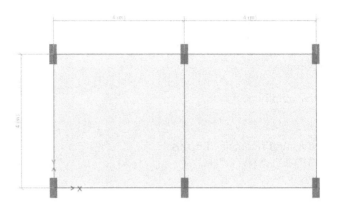

FIGURE 12.10: Plan

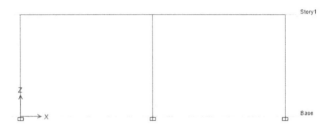

FIGURE 12.11: Elevation

- **Load:**
    - Load on roof: 1.5 kN/m^2.
    - Super dead load: 1 kN/m^2.
- **Concrete Properties:**
- **Density of Concrete:** 25 kN/m^3.
- **Density of infill:** 20 kN/m^3.
- **Concrete Grade:** M-20.
- **Plan Details:**
    - Length of plan: 8 m.
    - Breadth of plan: 4 m.
    - Number of columns: 6.
    - Grids along X-axis: 3.
    - Grids along Y-axis: 2.
- **Type of soil:** Hard.
- **Zone Type:** IV.
- **Building Type:** OMRF.

Python Program:
```
import numpy as np
import sympy as sy
```

Miscellaneous Aspects

```python
import numpy as np
import pandas as pd
import matplotlib.pyplot as plt
from numpy import matrix
from numpy import linalg as LA

B = float(input("Breadth of beam b: "))
D = float(input("Depth of beam d: "))
b = float(input("Width of column b: "))
d = float(input("Depth of column d: "))
S= float(input("Slab thickness S: "))
LL=float(input("Live load LL: "))
FF=float(input("floor finish load FL: "))
Dc=float(input("Density of concrete: "))
Dw=float(input("Density of brick: "))
tw=float(input("Thickness of wall: "))
fck=float(input("Grade of concrete fck: "))
L=float(input("Height of each storey: "))
Lp=float(input("Length of plan Lp: "))
Bp=float(input("Breadth of plan Bp: "))
N=float(input("Number of columns: "))
Nx=float(input("Number of grids along x: "))
Ny=float(input("Number of grids along y: "))
Numberoffloor=float(input("Number of storey: "))
#modulus of elasticity
E = (5000*(fck)**0.5)
I = (B*D**3)/12
EI = E*I
print("EI",EI)
#Stiffness of each storey
k= (12*EI)/(L**3)*3
print ("k",k)
#slab load
SL= Dc*S*Lp*Bp
print("SL",SL)
#self weight of column and beam
SWc= Dc*b*d*(L-D)*N
print ("SWc",SWc)
SWb= Dc*B*D*((Lp*Nx)+(Bp*Ny))
print ("SWb",SWb)
Wl= Dw*tw*(L-D)*((Lp*Nx)+(Bp*Ny))
#Imposed load
if LL<=3:
    IL=0.25*LL
```

```
else:
  IL=0.5*LL
print("IL",IL)
#Load on roof
m=(SL+(SWc/2)+SWb)
print("m",m)
Wn=(k/m)**0.5
print('Wn',Wn)
f=(2*3.142)/Wn
print('f',f)
```

Problem 6: (Figs. 12.12 and 12.13).

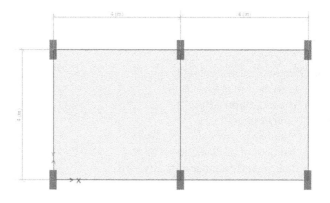

FIGURE 12.12: Plan

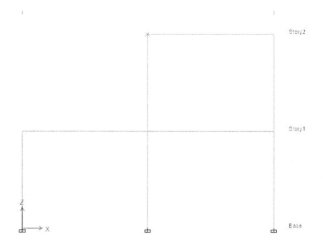

FIGURE 12.13: Elevation

Miscellaneous Aspects

The following data are used in the analysis:

- **Section Properties:**
    - Size of beam: 230×450 mm.
    - Size of column: 230×450 mm.
    - Depth of slab: 150 mm.
    - Thickness of wall: 150 mm.
- **Storey Properties:**
    - Height between the floors: 3.5 m.
    - Number of storey: 2.
- **Load:**
    - Live load on each floors: 3 kN/m^2.
    - Load on roof: 1.5 kN/m^2.
    - Super dead load: 1 kN/m^2.
- **Concrete Properties:**
- **Density of Concrete:** 25 kN/m^3.
- **Density of infill:** 20 kN/m^3.
- **Concrete Grade:** M-20.
- **Plan Details:**
    - Length of plan: 8 m.
    - Breadth of plan: 4 m.
    - Number of columns: 6.
    - Grids along X-axis: 3.
    - Grids along Y-axis: 2.
- **Type of soil:** Hard.
- **Zone Type:** IV.
- **Building Type:** OMRF.

Python program:
```
import numpy as np
import sympy as sy
import numpy as np
import pandas as pd
import matplotlib.pyplot as plt
from numpy import matrix
from numpy import linalg as LA
B = float(input("Breadth of beam b: "))
D = float(input("Depth of beam d: "))
b = float(input("Width of column b: "))
d = float(input("Depth of column d: "))
S= float(input("Slab thickness S: "))
LL=float(input("Live load LL: "))
FF=float(input("floor finish load FL: "))
Dc=float(input("Density of concrete: "))
Dw=float(input("Density of brick: "))
```

```
tw=float(input("Thickness of wall: "))
fck=float(input("Grade of concrete fck: "))
L=float(input("Height of each storey: "))
Lp=float(input("Length of plan Lp: "))
Bp=float(input("Breadth of plan Bp: "))
N=float(input("Number of columns: "))
Nx=float(input("Number of grids along x: "))
Ny=float(input("Number of grids along y: "))
Numberoffloor=float(input("Number of storey: "))
#modulus of elasticity
E = (5000*(fck)**0.5)
I = (B*D**3)/12
EI = E*I
print("EI",EI)
#Stiffness of each storey
k= (12*EI)/(L**3)*3
k1=(12*EI)/(L**3)*2
#slab load
SL= Dc*S*Lp*Bp
print("SL",SL)
#self weight of column and beam
SWc= Dc*b*d*(L-D)*N
print ("SWc",SWc)
SWb= Dc*B*D*((Lp*Nx)+(Bp*Ny))
print ("SWb",SWb)
Wl= Dw*tw*(L-D)*((Lp*Nx)+(Bp*Ny))
#Imposed load
if LL<=3:
   IL=0.25*LL
else:
   IL=0.5*LL
print("IL",IL)
#Load on each storey and roof
m=(SWc+SL+SWb+Wl+(IL*Lp*Bp))
m1=(SL+(SWc/2)+SWb)
print("m",m)
matrix1= np.matrix([[(k+k1),-k],[-k,k]])
matrix2=np.matrix([[(m),0],[0,(m1)]])
print("matrix2:",matrix2)

matrix3= matrix2.I
matrix4= matrix3*matrix1
eigenvalues, eigenvectors = np.linalg.eig(matrix4)
z = eigenvalues
```

Miscellaneous Aspects 251

```
for x in np.nditer(z, op_flags = ['readwrite']):
    x[...] = x**0.5
x1=z[0:1]
x2=z[1:2]
T1=2*3.142/x1
T2=2*3.142/x2
print("Time period=",T1,T2)
print("Frequency=",z)
```

Problem 7: (Figs. 12.14 and 12.15).

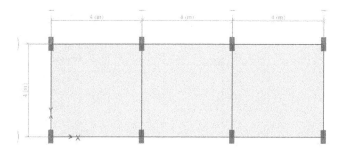

FIGURE 12.14: Plan

The following data are used in the analysis:

- **Section Properties:**
    - Size of beam: 230×450 mm.
    - Size of column: 230×600 mm.
    - Depth of Slab: 150 mm.
    - Thickness of wall: 150 mm.
- **Storey Properties:**
    - Height between the floors: 3.5 m.
    - Number of storey: 10.
- **Load:**
    - Live load on each floors: 3 kN/m^2.
    - Load on roof: 1.5 kN/m^2.
    - Super dead load: 1 kN/m^2.
- **Concrete Properties:**
- **Density of Concrete:** 25 kN/m^3.
- **Density of infill:** 20 kN/m^3.
- **Concrete Grade:** M-20.
- **Plan Details:**
    - Length of plan: 12 m.
    - Breadth of plan: 4 m.

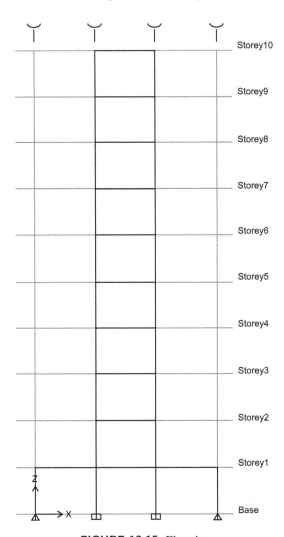

FIGURE 12.15: Elevation

- Number of columns: 8.
- Grids along X-axis: 3.
- Grids along Y-axis: 5.
- **Type of soil:** Hard.
- **Zone Type:** IV.
- **Building Type:** SMRF.

Python program:
import numpy as np
import sympy as sy

Miscellaneous Aspects

```python
import numpy as np
import pandas as pd
import matplotlib.pyplot as plt
from numpy import matrix
from numpy import linalg as LA

B = float(input("Breadth of beam b: "))
D = float(input("Depth of beam d: "))
b = float(input("Width of column b: "))
d = float(input("Depth of column d: "))
S= float(input("Slab thickness S: "))
LL=float(input("Live load LL: "))
FF=float(input("floor finish load FL: "))
Dc=float(input("Density of concrete: "))
Dw=float(input("Density of brick: "))
tw=float(input("Thickness of wall: "))
fck=float(input("Grade of concrete fck: "))
L=float(input("Height of each storey: "))
Lp=float(input("Length of plan Lp: "))
Bp=float(input("Breadth of plan Bp: "))
N=float(input("Number of columns: "))
Nx=float(input("Number of grids along x: "))
Ny=float(input("Number of grids along y: "))
Numberoffloor=float(input("Number of storey: "))
#modulus of elasticity
E = (5000*(fck)**(1/2))*10**3
I = (B*D**3)/12
EI = E*I
print("EI",EI)
#Stiffness of each store
k= 12*EI*N/L**3
print ("k",k)
#slab load
SL= Dc*S*Lp*Bp
print("SL",SL)
#self weight of column and beam
SWc= Dc*b*d*(L-D)*N
print ("SWc",SWc)
SWb= Dc*B*D*((Lp*Nx)+(Bp*Ny))
print ("SWb",SWb)
Wl= Dw*tw*(L-D)*((Lp*Nx)+(Bp*Ny))
#Imposed load
if LL<=3:

    IL=0.25*LL
```

else:

   IL=0.5*LL

print("IL",IL)
#Load on each storey and roof
m=(SWc+SL+SWb+Wl+(IL*Lp*Bp))
print("m",m)
m1=(SL+(SWc/2)+SWb)
print("m1",m1)
# [k]-[m]Y
matrix1= np.matrix([[(k1+k),-k,0,0,0,0,0,0,0,0],
    [-k,(k+k),-k,0,0,0,0,0,0,0],
    [0,-k,(k+k),-k,0,0,0,0,0,0],
    [0,0,-k,(k+k),-k,0,0,0,0,0],
    [0,0,0,-k,(k+k),-k,0,0,0,0],
    [0,0,0,0,-k,(k+k),-k,0,0,0],
    [0,0,0,0,0,-k,(k+k),-k,0,0],
    [0,0,0,0,0,0,-k,(k+k),-k,0],
    [0,0,0,0,0,0,0,-k,(k+k),-k],
    [0,0,0,0,0,0,0,0,-k,k]])
matrix2=np.matrix([[(m0),0,0,0,0,0,0,0,0,0],
    [0,(m),0,0,0,0,0,0,0,0],
    [0,0,(m),0,0,0,0,0,0,0],
    [0,0,0,(m),0,0,0,0,0,0],
    [0,0,0,0,(m),0,0,0,0,0],
    [0,0,0,0,0,(m),0,0,0,0],
    [0,0,0,0,0,0,(m),0,0,0],
    [0,0,0,0,0,0,0,(m),0,0],
    [0,0,0,0,0,0,0,0,(m),0],
    [0,0,0,0,0,0,0,0,0,(m1)]])
print("matrix2:",matrix2)
matrix3= matrix2.I
print("matrix3:",matrix3)
matrix4= matrix3*matrix1
print(matrix4)
eigenvalues, eigenvectors = np.linalg.eig(matrix4)
eigenvalues= np.linalg.eig(matrix4)
print ("Eigen values", eigenvalues)
print ("Eigen vectors", eigenvectors)
z = eigenvalues
for x in np.nditer(z, op_flags = ['readwrite']):
x[...] = x**0.5
x1=z[0:1]

Miscellaneous Aspects

```
x2=z[1:2]
x3=z[2:3]
x4=z[3:4]
x5=z[4:5]
x6=z[5:6]
x7=z[6:7]
x8=z[7:8]
x9=z[9:10]
x10=z[8:9]
print ("x9",x9)
print ("frequency:", z)
#k-w^2*m
#for x10:
i1=1
i2=(((k1+k)-(m0*x1**2))*i1)/k
i3=((-((2*k)-(m*x1**2))*i2)+(k*i1))/(-k)
i4=((-((2*k)-(m*x1**2))*i3)+(k*i2))/(-k)
i5=((-((2*k)-(m*x1**2))*i4)+(k*i3))/(-k)
i6=((-((2*k)-(m*x1**2))*i5)+(k*i4))/(-k)
i7=((-((2*k)-(m*x1**2))*i6)+(k*i5))/(-k)
i8=((-((2*k)-(m*x1**2))*i7)+(k*i6))/(-k)
i9=((-((2*k)-(m*x1**2))*i8)+(k*i7))/(-k)
i10=((-((2*k)-(m1*x1**2))*i9)+(k*i8))/(-k)
matrix6= np.matrix([[i1],[i2],[i3],[i4],[i5],[i6],[i7],[i8],[i9],[i10]])
mat1=matrix6.T
print("i2",i2)
u11=mat1*matrix2*matrix6
for y in np.nditer(u11, op_flags = ['readwrite']):

y[...] = y**0.5

y1=u11[0:1]
print(y1)
u1=(matrix6/(y1))
print("u1",u1)
#for x9
i21=1
i22=(((k1+k)-(m0*x2**2))*i21)/k
i23=((-((2*k)-(m*x2**2))*i22)+(k*i21))/(-k)
i24=((-((2*k)-(m*x2**2))*i23)+(k*i22))/(-k)
i25=((-((2*k)-(m*x2**2))*i24)+(k*i23))/(-k)
i26=((-((2*k)-(m*x2**2))*i25)+(k*i24))/(-k)
i27=((-((2*k)-(m*x2**2))*i26)+(k*i25))/(-k)
i28=((-((2*k)-(m*x2**2))*i27)+(k*i26))/(-k)
```

```
i29=((-((2*k)-(m*x2**2))*i28)+(k*i27))/(-k)
i210=((-((2*k)-(m1*x2**2))*i29)+(k*i28))/(-k)
matrix7= np.matrix([[i21],[i22],[i23],[i24],[i25],[i26],[i27],[i28],[i29],[i210]])
mat2=matrix7.T
u12=mat2*matrix2*matrix7
for y in np.nditer(u12, op_flags = ['readwrite']):

y[...] = y**0.5

y2=u12[0:1]
print(y2)
u2=(matrix7/(y2))
print("u2",u2)

#for x8
i31=1
i32=(((k1+k)-(m0*x3**2))*i31)/k
i33=((-((2*k)-(m*x3**2))*i32)+(k*i31))/(-k)
i34=((-((2*k)-(m*x3**2))*i33)+(k*i32))/(-k)
i35=((-((2*k)-(m*x3**2))*i34)+(k*i33))/(-k)
i36=((-((2*k)-(m*x3**2))*i35)+(k*i34))/(-k)
i37=((-((2*k)-(m*x3**2))*i36)+(k*i35))/(-k)
i38=((-((2*k)-(m*x3**2))*i37)+(k*i36))/(-k)
i39=((-((2*k)-(m*x3**2))*i38)+(k*i37))/(-k)
i310=((-((2*k)-(m1*x3**2))*i39)+(k*i38))/(-k)
matrix8= np.matrix([[i31],[i32],[i33],[i34],[i35],[i36],[i37],[i38],[i39],[i310]])
mat3=matrix8.T
u13=mat3*matrix2*matrix8
for y in np.nditer(u13, op_flags = ['readwrite']):

y[...] = y**0.5

y3=u13[0:1]
print(y3)
u3=(matrix8/(y3))
print("u3",u3)

#for x7
i41=1
i42=(((k1+k)-(m0*x4**2))*i41)/k
i43=((-((2*k)-(m*x4**2))*i42)+(k*i41))/(-k)
i44=((-((2*k)-(m*x4**2))*i43)+(k*i42))/(-k)
i45=((-((2*k)-(m*x4**2))*i44)+(k*i43))/(-k)
i46=((-((2*k)-(m*x4**2))*i45)+(k*i44))/(-k)
```

Miscellaneous Aspects 257

```
i47=((-((2*k)-(m*x4**2))*i46)+(k*i45))/(-k)
i48=((-((2*k)-(m*x4**2))*i47)+(k*i46))/(-k)
i49=((-((2*k)-(m*x4**2))*i48)+(k*i47))/(-k)
i410=((-((2*k)-(m1*x4**2))*i49)+(k*i48))/(-k)
matrix9= np.matrix([[i41],[i42],[i43],[i44],[i45],[i46],[i47],[i48],[i49],[i410]])
mat4=matrix9.T
u14=mat4*matrix2*matrix9
for y in np.nditer(u14, op_flags = ['readwrite']):

y[...] = y**0.5

y4=u14[0:1]
print(y4)
u4=(matrix9/(y4))
print("u4",u4)

#for x6
i51=1
i52=(((k1+k)-(m0*x5**2))*i51)/k
i53=((-((2*k)-(m*x5**2))*i52)+(k*i51))/(-k)
i54=((-((2*k)-(m*x5**2))*i53)+(k*i52))/(-k)
i55=((-((2*k)-(m*x5**2))*i54)+(k*i53))/(-k)
i56=((-((2*k)-(m*x5**2))*i55)+(k*i54))/(-k)
i57=((-((2*k)-(m*x5**2))*i56)+(k*i55))/(-k)
i58=((-((2*k)-(m*x5**2))*i57)+(k*i56))/(-k)
i59=((-((2*k)-(m*x5**2))*i58)+(k*i57))/(-k)
i510=((-((2*k)-(m1*x5**2))*i59)+(k*i58))/(-k)
matrix10= np.matrix([[i51],[i52],[i53],[i54],[i55],[i56],[i57],[i58],[i59],[i510]])
mat5=matrix10.T
u15=mat5*matrix2*matrix10
for y in np.nditer(u15, op_flags = ['readwrite']):

y[...] = y**0.5

y5=u15[0:1]
print(y5)
u5=(matrix10/(y5))
print("u5",u5)

#for x5
i61=1
i62=(((k1+k)-(m0*x6**2))*i61)/k
i63=((-((2*k)-(m*x6**2))*i62)+(k*i61))/(-k)
i64=((-((2*k)-(m*x6**2))*i63)+(k*i62))/(-k)
```

```
i65=((-((2*k)-(m*x6**2))*i64)+(k*i63))/(-k)
i66=((-((2*k)-(m*x6**2))*i65)+(k*i64))/(-k)
i67=((-((2*k)-(m*x6**2))*i66)+(k*i65))/(-k)
i68=((-((2*k)-(m*x6**2))*i67)+(k*i66))/(-k)
i69=((-((2*k)-(m*x6**2))*i68)+(k*i67))/(-k)
i610=((-((2*k)-(m1*x6**2))*i69)+(k*i68))/(-k)
matrix11= np.matrix([[i61],[i62],[i63],[i64],[i65],[i66],[i67],[i68],[i69],[i610]])
mat6=matrix11.T
u16=mat6*matrix2*matrix11
for y in np.nditer(u16, op_flags = ['readwrite']):

y[...] = y**0.5

y6=u16[0:1]
print(y6)
u6=(matrix11/(y6))
print("u6",u6)

#for x4
i71=1
i72=(((k1+k)-(m0*x7**2))*i71)/k
i73=((-((2*k)-(m*x7**2))*i72)+(k*i71))/(-k)
i74=((-((2*k)-(m*x7**2))*i73)+(k*i72))/(-k)
i75=((-((2*k)-(m*x7**2))*i74)+(k*i73))/(-k)
i76=((-((2*k)-(m*x7**2))*i75)+(k*i74))/(-k)
i77=((-((2*k)-(m*x7**2))*i76)+(k*i75))/(-k)
i78=((-((2*k)-(m*x7**2))*i77)+(k*i76))/(-k)
i79=((-((2*k)-(m*x7**2))*i78)+(k*i77))/(-k)
i710=((-((2*k)-(m1*x7**2))*i79)+(k*i78))/(-k)
matrix12= np.matrix([[i71],[i72],[i73],[i74],[i75],[i76],[i77],[i78],[i79],[i710]])
mat7=matrix12.T
u17=mat7*matrix2*matrix12
for y in np.nditer(u17, op_flags = ['readwrite']):

y[...] = y**0.5

y7=u17[0:1]
print(y7)
u7=(matrix12/(y7))
print("u7",u7)

#for x3
i81=1
i82=(((k1+k)-(m0*x8**2))*i81)/k
```

Miscellaneous Aspects 259

```
i83=((-((2*k)-(m*x8**2))*i82)+(k*i81))/(-k)
i84=((-((2*k)-(m*x8**2))*i83)+(k*i82))/(-k)
i85=((-((2*k)-(m*x8**2))*i84)+(k*i83))/(-k)
i86=((-((2*k)-(m*x8**2))*i85)+(k*i84))/(-k)
i87=((-((2*k)-(m*x8**2))*i86)+(k*i85))/(-k)
i88=((-((2*k)-(m*x8**2))*i87)+(k*i86))/(-k)
i89=((-((2*k)-(m*x8**2))*i88)+(k*i87))/(-k)
i810=((-((2*k)-(m1*x8**2))*i89)+(k*i88))/(-k)
matrix13= np.matrix([[i81],[i82],[i83],[i84],[i85],[i86],[i87],[i88],[i89],[i810]])
mat8=matrix13.T
u18=mat8*matrix2*matrix13
for y in np.nditer(u18, op_flags = ['readwrite']):

y[...] = y**0.5


y8=u18[0:1]
print(y8)
u8=(matrix13/(y8))
print("u8",u8)

#for x2
i91=1
i92=(((k1+k)-(m0*x9**2))*i91)/k
i93=((-((2*k)-(m*x9**2))*i92)+(k*i91))/(-k)
i94=((-((2*k)-(m*x9**2))*i93)+(k*i92))/(-k)
i95=((-((2*k)-(m*x9**2))*i94)+(k*i93))/(-k)
i96=((-((2*k)-(m*x9**2))*i95)+(k*i94))/(-k)
i97=((-((2*k)-(m*x9**2))*i96)+(k*i95))/(-k)
i98=((-((2*k)-(m*x9**2))*i97)+(k*i96))/(-k)
i99=((-((2*k)-(m*x9**2))*i98)+(k*i97))/(-k)
i910=((-((2*k)-(m1*x9**2))*i99)+(k*i98))/(-k)
matrix14= np.matrix([[i91],[i92],[i93],[i94],[i95],[i96],[i97],[i98],[i99],[i910]])
mat9=matrix14.T
u19=mat9*matrix2*matrix14
for y in np.nditer(u19, op_flags = ['readwrite']):

y[...] = y**0.5


y9=u19[0:1]
print(y9)
u9=(matrix14/(y9))
print("u9",u9)

#for x1
i01=1
```

```
i02=(((k1+k)-(m0*x10**2))*i01)/k
i03=((-((2*k)-(m*x10**2))*i02)+(k*i01))/(-k)
i04=((-((2*k)-(m*x10**2))*i03)+(k*i02))/(-k)
i05=((-((2*k)-(m*x10**2))*i04)+(k*i03))/(-k)
i06=((-((2*k)-(m*x10**2))*i05)+(k*i04))/(-k)
i07=((-((2*k)-(m*x10**2))*i06)+(k*i05))/(-k)
i08=((-((2*k)-(m*x10**2))*i07)+(k*i06))/(-k)
i09=((-((2*k)-(m*x10**2))*i08)+(k*i07))/(-k)
i010=((-((2*k)-(m1*x10**2))*i09)+(k*i08))/(-k)
matrix15= np.matrix([[i01],[i02],[i03],[i04],[i05],[i06],[i07],[i08],[i09],[i010]])
mat10=matrix15.T
u20=mat10*matrix2*matrix15
for y in np.nditer(u20, op_flags = ['readwrite']):

y[...] = y**0.5

y10=u20[0:1]
print(y10)
u10=(matrix15/(y10))
print("u10",u10)


T1=2*3.142/x1
T2=2*3.142/x2
T3=2*3.142/x3
T4=2*3.142/x4
T5=2*3.142/x5
T6=2*3.142/x6
T7=2*3.142/x7
T8=2*3.142/x8
T9=2*3.142/x9
T10=2*3.142/x10
print ("T1",T1)
df1 = pd.DataFrame(np.array(u1), columns=['X'])
df2 = pd.DataFrame(np.array(u2), columns=['b'])
df3 = pd.DataFrame(np.array(u3), columns=['c'])
df4 = pd.DataFrame(np.array(u4), columns=['d'])
df5 = pd.DataFrame(np.array(u5), columns=['e'])
df6 = pd.DataFrame(np.array(u6), columns=['f'])
df7 = pd.DataFrame(np.array(u7), columns=['g'])
df8 = pd.DataFrame(np.array(u8), columns=['h'])
df9 = pd.DataFrame(np.array(u9), columns=['i'])
df10 = pd.DataFrame(np.array(u10), columns=['j'])
X= df1.X
```

Miscellaneous Aspects

```
b = df2.b
c = df3.c
d = df4.d
e = df5.e
f = df6.f
g = df7.g
h = df8.h
i = df9.i
j= df10.j
print("X",X)
print("X[1]",X[1])
#Modal Participation Factor:
p1=sum(m0*X)/sum(m0*(X**2))
p2=sum(m*b)/sum(m*(b**2))
p3=sum(m*c)/sum(m*(c**2))
p4=sum(m*d)/sum(m*(d**2))
p5=sum(m*e)/sum(m*(e**2))
p6=sum(m*f)/sum(m*(f**2))
p7=sum(m*g)/sum(m*(g**2))
p8=sum(m*h)/sum(m*(h**2))
p9=sum(m*i)/sum(m*(i**2))
p10=sum(m1*j)/sum(m1*(j**2))
print ("p1",p1,p2,p3,p4,p5,p6,p7,p8,p9,p10)

#Modal Mass:
g1=9.81
M1=sum(m0*X)**2/sum((m0*(X**2))*g1)
M2=sum(m*b)**2/sum((m*(b**2))*g1)
M3=sum(m*c)**2/sum((m*(c**2))*g1)
M4=sum(m*d)**2/sum((m*(d**2))*g1)
M5=sum(m*e)**2/sum((m*(e**2))*g1)
M6=sum(m*f)**2/sum((m*(f**2))*g1)
M7=sum(m*g)**2/sum((m*(g**2))*g1)
M8=sum(m*h)**2/sum((m*(h**2))*g1)
M9=sum(m*i)**2/sum((m*(i**2))*g1)
M10=sum(m1*j)**2/sum((m1*(j**2))*g1)
print ("M10",M1,M2,M3,M4,M5,M6,M7,M8,M9,M10)
Z=0.24
I=1
R=5
T=T1
if 0<=T<=0.1:
   (s1)=1+15*T
elif 0.1<=T<=0.40:
```

```
    (s1)=2.50
else:
    0.40<=T<=4
    (s1)=1/T
print("s1",s1)

T=T2
if 0<=T<=0.1:
    (s2)=1+15*T
elif 0.1<=T<=0.40:
    (s2)=2.50
else:
    0.40<=T<=4
    (s2)=1/T
print("s2",s2)
T=T3
if 0<=T<=0.1:
    (s3)=1+15*T
elif 0.1<=T<=0.40:
    (s3)=2.50
else:
    0.40<=T<=4
    (s3)=1/T
print("s3",s3)
T=T4
if 0<=T<=0.1:
    (s4)=1+15*T
elif 0.1<=T<=0.40:
    (s4)=2.50
else:
    0.40<=T<=4
    (s4)=1/T
print("s4",s4)
T=T5

if 0<=T<=0.1:
    (s5)=1+15*T
elif 0.1<=T<=0.40:
    (s5)=2.50
else:
    0.40<=T<=4
    (s5)=1/T
print("s5",s5)
T=T6
```

Miscellaneous Aspects 263

```
if 0<=T<=0.1:
    (s6)=1+15*T
elif 0.1<=T<=0.40:
    (s6)=2.50
else:
    0.40<=T<=4
    (s6)=1/T
print("s6",s6)
T=T7
if 0<=T<=0.1:
    (s7)=1+15*T

elif 0.1<=T<=0.40:
    (s7)=2.50
else:
    0.40<=T<=4
    (s7)=1/T
print("s7",s7)
T=T8
if 0<=T<=0.1:
    (s8)=1+15*T
elif 0.1<=T<=0.40:
    (s8)=2.50
else:
    0.40<=T<=4
    (s8)=1/T
print("s8",s8)
T=T9
if 0<=T<=0.1:
    (s9)=1+15*T
elif 0.1<=T<=0.40:
    (s9)=2.50
else:
    0.40<=T<=4
    (s9)=1/T
print("s9",s9)
T=T10
if 0<=T<=0.1:
    (s10)=1+15*T
elif 0.1<=T<=0.40:
    (s10)=2.50
else:
    0.40<=T<=4
    (s10)=1/T
```

```
print("s10",s10)
#T=((2*3.142)/x)
g1=9.81
SaA=s1
SaB=s2
SaC=s3
SaD=s4
SaE=s5
SaF=s6
SaG=s7
SaH=s8
SaI=s9
SaJ=s10
a1=(Z/2)*(I/R)*(SaA)
print("a1",a1)
a2=(Z/2)*(I/R)*(SaB)
a3=(Z/2)*(I/R)*(SaC)
a4=(Z/2)*(I/R)*(SaD)
a5=(Z/2)*(I/R)*(SaE)
a6=(Z/2)*(I/R)*(SaF)
a7=(Z/2)*(I/R)*(SaG)
a8=(Z/2)*(I/R)*(SaH)
a9=(Z/2)*(I/R)*(SaI)
a10=(Z/2)*(I/R)*(SaJ)
print("a8",a8)
print("a9",a9)
print("X",X)
#at first floor
Q1=([a1*p1*m0*g1*X])
Q11=(Q1[1:10])
Q12=(Q1[2:10])
Q13=(Q1[3:10])
Q14=(Q1[4:10])
Q15=(Q1[5:10])
Q16=(Q1[6:10])
Q17=(Q1[7:10])
Q18=(Q1[8:10])
Q19=(Q1[9:10])
print ("Q1",Q1,Q11)

V11=sum(Q1)
V12=sum(Q11)
V13=sum(Q12)
V14=sum(Q13)
```

Miscellaneous Aspects

V15=sum(Q14)
V16=sum(Q15)
V17=sum(Q16)
V18=sum(Q17)
V19=sum(Q18)
V20=sum(Q19)
print("V11",V11)
print("V12",V12)
Q2=([a2*p2*m*g1*b])
Q21=(Q2[1:10])
Q22=(Q2[2:10])
Q23=(Q2[3:10])
Q24=(Q2[4:10])
Q25=(Q2[5:10])
Q26=(Q2[6:10])
Q27=(Q2[7:10])
Q28=(Q2[8:10])
Q29=(Q2[9:10])
print ("Q2",Q2)

V21=sum(Q2)
V22=sum(Q21)
V23=sum(Q22)
V24=sum(Q23)
V25=sum(Q24)
V26=sum(Q25)
V27=sum(Q26)
V28=sum(Q27)
V29=sum(Q28)
V30=sum(Q29)

Q3=([a3*p3*m*g1*c])
Q31=(Q3[1:10])
Q32=(Q3[2:10])
Q33=(Q3[3:10])
Q34=(Q3[4:10])
Q35=(Q3[5:10])
Q36=(Q3[6:10])
Q37=(Q3[7:10])
Q38=(Q3[8:10])
Q39=(Q3[9:10])

V31=sum(Q3)
V32=sum(Q31)

V33=sum(Q32)
V34=sum(Q33)
V35=sum(Q34)
V36=sum(Q35)
V37=sum(Q36)
V38=sum(Q37)
V39=sum(Q38)
V40=sum(Q39)

Q4=([a4*p4*m*g1*d])
Q41=(Q4[1:10])
Q42=(Q4[2:10])
Q43=(Q4[3:10])
Q44=(Q4[4:10])
Q45=(Q4[5:10])
Q46=(Q4[6:10])
Q47=(Q4[7:10])
Q48=(Q4[8:10])
Q49=(Q4[9:10])

V41=sum(Q4)
V42=sum(Q41)
V43=sum(Q42)
V44=sum(Q43)
V45=sum(Q44)
V46=sum(Q45)
V47=sum(Q46)
V48=sum(Q47)
V49=sum(Q48)
V50=sum(Q49)
print("V41",V41)
Q5=([a5*p5*m*g1*e])
Q51=(Q5[1:10])
Q52=(Q5[2:10])
Q53=(Q5[3:10])
Q54=(Q5[4:10])
Q55=(Q5[5:10])
Q56=(Q5[6:10])
Q57=(Q5[7:10])
Q58=(Q5[8:10])
Q59=(Q5[9:10])

V51=sum(Q5)
V52=sum(Q51)

Miscellaneous Aspects 267

V53=sum(Q52)
V54=sum(Q53)
V55=sum(Q54)
V56=sum(Q55)
V57=sum(Q56)
V58=sum(Q57)
V59=sum(Q58)
V60=sum(Q59)

Q6=([a6*p6*m*g1*f])
Q61=(Q6[1:10])
Q62=(Q6[2:10])
Q63=(Q6[3:10])
Q64=(Q6[4:10])
Q65=(Q6[5:10])
Q66=(Q6[6:10])
Q67=(Q6[7:10])
Q68=(Q6[8:10])
Q69=(Q6[9:10])

Q7=([a7*p7*m*g1*g])
Q71=(Q7[1:10])
Q72=(Q7[2:10])
Q73=(Q7[3:10])
Q74=(Q7[4:10])
Q75=(Q7[5:10])
Q76=(Q7[6:10])
Q77=(Q7[7:10])
Q78=(Q7[8:10])
Q79=(Q7[9:10])
print ("Q7",Q7)

Q8=([a8*p8*m*g1*h])
Q81=(Q8[1:10])
Q82=(Q8[2:10])
Q83=(Q8[3:10])
Q84=(Q8[4:10])
Q85=(Q8[5:10])
Q86=(Q8[6:10])
Q87=(Q8[7:10])
Q88=(Q8[8:10])
Q89=(Q8[9:10])
print ("Q8",Q8)

Q9=([a9*p9*m*g1*i])
Q91=(Q9[1:10])
Q92=(Q9[2:10])
Q93=(Q9[3:10])
Q94=(Q9[4:10])
Q95=(Q9[5:10])
Q96=(Q9[6:10])
Q97=(Q9[7:10])
Q98=(Q9[8:10])
Q99=(Q9[9:10])

Q10=([a10*p10*m1*g1*j])
Q01=(Q10[1:10])
Q02=(Q10[2:10])
Q03=(Q10[3:10])
Q04=(Q10[4:10])
Q05=(Q10[5:10])
Q06=(Q10[6:10])
Q07=(Q10[7:10])
Q08=(Q10[8:10])
Q09=(Q10[9:10])

V61=sum(Q6)
V62=sum(Q61)
V63=sum(Q62)
V64=sum(Q63)
V65=sum(Q64)
V66=sum(Q65)
V67=sum(Q66)
V68=sum(Q67)
V69=sum(Q68)
V70=sum(Q69)

V71=sum(Q7)
V72=sum(Q71)
V73=sum(Q72)
V74=sum(Q73)
V75=sum(Q74)
V76=sum(Q75)
V77=sum(Q76)
V78=sum(Q77)
V79=sum(Q78)
V80=sum(Q79)
V81=sum(Q8)

Miscellaneous Aspects 269

```
V82=sum(Q81)
V83=sum(Q82)
V84=sum(Q83)
V85=sum(Q84)
V86=sum(Q85)
V87=sum(Q86)
V88=sum(Q87)
V89=sum(Q88)
V90=sum(Q89)

V91=sum(Q9)
V92=sum(Q91)
V93=sum(Q92)
V94=sum(Q93)
V95=sum(Q94)
V96=sum(Q95)
V97=sum(Q96)
V98=sum(Q97)
V99=sum(Q98)
V100=sum(Q99)

V01=sum(Q10)
V02=sum(Q01)
V03=sum(Q02)
V04=sum(Q03)
V05=sum(Q04)
V06=sum(Q05)
V07=sum(Q06)
V08=sum(Q07)
V09=sum(Q08)
V101=sum(Q09)
print("V101",V101)

V1=(V11**2+V21**2+V31**2+V41**2+V51**2+V61**2+V71**2+V81**2
+V91**2+V01**2)**0.5
V2=(V12**2+V22**2+V32**2+V42**2+V52**2+V62**2+V72**2+V82**2
+V92**2+V02**2)**0.5
V3=(V13**2+V23**2+V33**2+V43**2+V53**2+V63**2+V73**2+V83**2
+V93**2+V03**2)**0.5
V4=(V14**2+V24**2+V34**2+V44**2+V54**2+V64**2+V74**2+V84**2
+V94**2+V04**2)**0.5
V5=(V15**2+V25**2+V35**2+V45**2+V55**2+V65**2+V75**2+V85**2
+V95**2+V05**2)**0.5
V6=(V16**2+V26**2+V36**2+V46**2+V56**2+V66**2+V76**2+V86**2
```

```
+V96**2+V06**2)**0.5
V7=(V17**2+V27**2+V37**2+V47**2+V57**2+V67**2+V77**2+V87**2
+V97**2+V07**2)**0.5
V8=(V18**2+V28**2+V38**2+V48**2+V58**2+V68**2+V78**2+V88**2
+V98**2+V08**2)**0.5
V9=(V19**2+V29**2+V39**2+V49**2+V59**2+V69**2+V79**2+V89**2
+V99**2+V09**2)**0.5
V10=(V20**2+V30**2+V40**2+V50**2+V60**2+V70**2+V80**2+V90**2
+V100**2+V101**2)**0.5
print("V1",V1)

#Lateral Forces at each Storey:
F10=V10
F9=V9-V10
F8=V8-V9
F7=V7-V8
F6=V6-V7
F5=V5-V6
F4=V4-V5
F3=V3-V4
F2=V2-V3
F1=V1-V2

print ("F1", F1)
print ("F2", F2)
print ("F3", F3)
print ("F4", F4)
print ("F5", F5)
print ("F6", F6)
print ("F7", F7)
print ("F8", F8)
print ("F9", F9)
print ("F10", F10)

x1 = [X[0],X[1],X[2],X[3],X[4],X[5],X[6],X[7],X[8],X[9]]
y = [1,2,3,4,5,6,7,8,9,10]
plt.plot(x1, y)
plt.xlabel('Displacement')
plt.ylabel('Storey')
plt.title('Mode shape 1')
plt.show()
x2 = [b[0],b[1],b[2],b[3],b[4],b[5],b[6],b[7],b[8],b[9]]
y = [1,2,3,4,5,6,7,8,9,10]
plt.plot(x2, y)
```

```
plt.xlabel('Displacement')
plt.ylabel('Storey')
plt.title('Mode shape 2')
plt.show()
x3 = [c[0],c[1],c[2],c[3],c[4],c[5],c[6],c[7],c[8],c[9]]
y = [1,2,3,4,5,6,7,8,9,10]
plt.plot(x3, y)
plt.xlabel('Displacement')
plt.ylabel('Storey')
plt.title('Mode shape 3')
plt.show()
x4 = [d[0],d[1],d[2],d[3],d[4],d[5],d[6],d[7],d[8],d[9]]
y = [1,2,3,4,5,6,7,8,9,10]
plt.plot(x4, y)
plt.xlabel('Displacement')
plt.ylabel('Storey')
plt.title('Mode shape 4')
plt.show()
x5 = [e[0],e[1],e[2],e[3],e[4],e[5],e[6],e[7],e[8],e[9]]
y = [1,2,3,4,5,6,7,8,9,10]
plt.plot(x5, y)
plt.xlabel('Displacement')
plt.ylabel('Storey')
plt.title('Mode shape 5')
plt.show()
x6 = [f[0],f[1],f[2],f[3],f[4],f[5],f[6],f[7],f[8],f[9]]
y = [1,2,3,4,5,6,7,8,9,10]
plt.plot(x6, y)
plt.xlabel('Displacement')
plt.ylabel('Storey')
plt.title('Mode shape 6')
plt.show()
x7 = [g[0],g[1],g[2],g[3],g[4],g[5],g[6],g[7],g[8],g[9]]
y = [1,2,3,4,5,6,7,8,9,10]
plt.plot(x7, y)
plt.xlabel('Displacement')
plt.ylabel('Storey')
plt.title('Mode shape 7')
plt.show()
x8 = [h[0],h[1],h[2],h[3],h[4],h[5],h[6],h[7],h[8],h[9]]
y = [1,2,3,4,5,6,7,8,9,10]
plt.plot(x8, y)
plt.xlabel('Displacement')
plt.ylabel('Storey')
```

```
plt.title('Mode shape 8')
plt.show()
x9 = [i[0],i[1],i[2],i[3],i[4],i[5],i[6],i[7],i[8],i[9]]
y = [1,2,3,4,5,6,7,8,9,10]
plt.plot(x9, y)
plt.xlabel('Displacement')
plt.ylabel('Storey')
plt.title('Mode shape 9')
plt.show()
x10 = [j[0],j[1],j[2],j[3],j[4],j[5],j[6],j[7],j[8],j[9]]
y = [1,2,3,4,5,6,7,8,9,10]
plt.plot(x10, y)
plt.xlabel('Displacement')
plt.ylabel('Storey')
plt.title('Mode shape 10')
plt.show()
print("Eigen values=",eigenvalues)
print("Frequency=",z)
print("Modal participation factor",p1,p2,p3,p4,p5,p6,p7,p8,p9,p10)
print("Modal mass",M1,M2,M3,M4,M5,M6,M7,M8,M9,M10)
print("Time period",T1,T2,T3,T4,T5,T6,T7,T8,T9,T10)
print("Mode shape",u1,u2,u3,u4,u5,u6,u7,u8,u9,u10)
```

Results
Problem 1:

Table 12.6 Natural frequency and natural time period

	Python	Manual
Natural frequency (rad/s)	20.22	20.22
Natural time period (seconds)	0.31	0.31

Problem 2:

Table 12.7 Natural frequency and natural time period

Natural frequency(rad/s)	ω_1	ω_2	Natural time period(sec)	T1	T2
Python	40.09	13.89	Python	0.16	0.45
Manual	40.09	13.88	Manual	0.16	0.45

Problem 3:

Table 12.8 Natural frequencies of different modes

Natural frequency(rad/s)	$\omega 1$	$\omega 2$	$\omega 3$	$\omega 4$	$\omega 5$	$\omega 6$	$\omega 7$	$\omega 8$	$\omega 9$	$\omega 10$
Python	2.78	8.25	13.42	18.09	22.12	25.54	28.51	30.96	32.68	34.04
ETABS	3.47	10.31	16.77	22.61	27.65	31.92	35.63	38.70	40.85	42.55
Manual	2.78	8.25	13.42	18.09	22.12	25.54	28.51	30.96	32.68	34.04

Table 12.9 Natural time period in seconds

Natural Time period	T1	T2	T3	T4	T5	T6	T7	T8	T9	T10
Python	2.25	0.76	0.46	0.34	0.28	0.24	0.22	0.20	0.18	0.19
ETABS	2.81	0.95	0.57	0.42	0.35	0.30	0.27	0.25	0.22	0.23
Manual	2.25	0.76	0.46	0.34	0.28	0.24	0.22	0.20	0.18	0.19

Table 12.10 Modal participation factor and modal mass

Mode	Modal participation factor			Modal mass		
	Python	ETABS	Manual	Python	ETABS	Manual
1	46.24	57.80	46.24	361.85	452.13	361.85
2	8.77	10.96	8.77	9.024	11.28	9.024
3	13.74	17.17	13.74	22.55	28.18	22.55
4	-0.39	-0.48	-0.39	0.018	0.022	0.018
5	10.11	12.63	10.11	11.45	14.31	11.45
6	-2.03	-2.53	-2.03	0.44	0.55	0.44
7	3.42	4.27	3.42	1.19	1.48	1.19
8	1.91	2.38	1.91	0.368	0.46	0.368
9	-4.01	5.01	-4.01	1.64	2.05	1.64
10	14.50	18.12	14.50	10.94	13.67	10.94

By adoption of the Python programming language, it is easy to compute complex problems with accuracy and precision.

In the study, the results from Python have been compared with those obtained from manual calculation as well as the well-known software ETABS. The results from Python and results from manual calculations agree very closely, while results from ETABS are farther away (Tables 12.6-12.11).

Table 12.11 Storey shear force and lateral forces at each storey level

Story	Story shear force (kN)			Lateral force at each story level(kN)		
	Python	ETABS	Manual	Python	ETABS	Manual
1	114.97	143.71	114.97	32.87	41.109	32.87
2	82.10	102.62	82.10	7.83	9.79	7.83
3	74.27	92.83	74.27	2.65	3.31	2.65
4	71.62	89.52	71.62	-8.76	-10.95	-8.76
5	80.38	100.47	80.38	14.24	17.81	14.24
6	66.14	82.67	66.14	12.94	16.17	12.94
7	53.20	66.50	53.20	-15.31	-19.13	-15.31
8	68.51	85.63	68.51	7.11	8.88	7.11
9	61.4	76.75	61.4	10.13	12.67	10.13
10	51.27	64.08	51.27	51.27	64.08	51.27

The present study has opened up a new avenue for earthquake analysis which is least expensive. It has shown that there is a wide scope of seismic analysis.

Because Python is an open source code, no expense is involved in its application.

In the present work, it has been amply demonstrated that commercial software such as ETABS and STAAD PRO, which are fairly expensive, let us say, and Python can be used in their place.

12.4　HYBRID BUILDING UNDER SEISMIC FORCES

12.4.1　INTRODUCTION

The following work is a part of the dissertation work of student Amir Khan. Minimal or nil use of cement in industry is the key to reduce carbon emission to the atmosphere. Of course, steel also is a high energy material, but when used as a structural member due to its large value of E, thinner sections and lesser quantity of material are possible. Besides, there are other advantages like the fact that it can be easily transported to greater height in a tall building. Concrete can also be pumped to greater heights, visual checking of quality of concrete poured at such heights, and also checking the reinforcement details are difficult, unless drones are used.

In view of the above, it will be prudent to use RC in lower storeys and steel in upper storeys. It is also observed in such buildings that the base shear is significantly less than that of RC structures. However, the main problem is that of connections between steel and RC sections.

12.4.2　TYPES OF CONNECTIONS

Some connections, namely simple, rigid and semi-rigid.

12.4.3 EARTHQUAKE RESPONSES OF HYBRID BUILDING

A $G+10$ building with various distributions of steel and RCC frame below or above having regular plan situated in seismic zone II having soil condition medium. Select an ISMB 450 steel frame and 450×300 RCC frames, respectively with a slab thickness 150 mm. The building is analyzed using ETABS (2016) as per as IS 1893 with response reduction factor, 5 and 5% damping. The base dimensions are 24×24 m as described in Fig. 12.16.

The following four models are studied:

Model 1- represents complete concrete building
Model 2- represents complete steel building
Model 3- represents 25% steel and 75% of concrete building (hybrid structure)
Model 4- represents 50% concrete and 50% of steel building (hybrid structure)

Results are shown in Tables 12.12 through Table 12.14.
1. Storey displacement
2. Modal time period
3. Base shear (equivalent static)

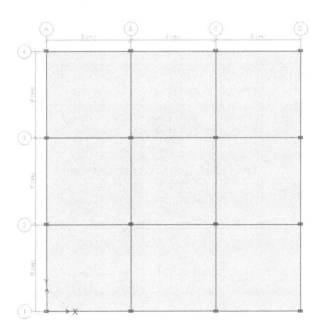

FIGURE 12.16: Sectional plan

Table 12.12 The maximum storey displacement

Type of model	Maximum displacement (mm)
Model 1	275.848
Model 2	203.04
Model 3	239.117
Model 4	225.3

Table 12.13 Modal period

Mode	Periods (sec)			
	Model 1	Model 2	Model 3	Model 4
1	3.302	5.59	3.467	4.404
2	2.825	2.71	2.567	2.436
3	2.617	2.48	2.389	2.355
4	1.089	2.271	1.758	1.707
5	0.919	1.295	0.897	1.188
6	0.858	0.929	0.873	0.924
7	0.639	0.883	0.862	0.855
8	0.527	0.813	0.708	0.817
9	0.499	0.729	0.583	0.637
10	0.446	0.622	0.521	0.553
11	0.357	0.549	0.505	0.508
12	0.343	0.505	0.431	0.498

Table 12.14 The maximum base shear

Type of model	Max base shear (kN)
Model 1	42683
Model 2	2251.0195
Model 3	2038.16
Model 4	1896.25

1. The maximum base shear is less in hybrid structure compared to that in a RC structure

2. The maximum displacement and drift are less in a hybrid structure compared to those in a steel structure

3. However, the steel structure has the advantage of large ductility

12.5 ANALYSIS AND DESIGN OF BLAST RESISTING STRUCTURES

12.5.1 GENERAL CHARACTERISTICS OF BLAST AND CONSEQUENCES ON STRUCTURES

The work reported here is a part of the dissertation work of student Sherlin Sheeba (IS 4991:1968). Because of an explosion, a shock wave is generated in the air that moves outwards in all directions from the point of burst with high velocity causing at all points time-dependent pressure and suction effects. The shock wave consists of an initial positive pressure phase followed by a negative phase at any point as illustrated in Figure 12.17. The shock wave is followed by blast wind causing dynamic pressure due to drag effects. Reflected pressure is caused instantaneously by wave diffraction at an obstructing surface that clears in time, depending on the extent to which the surface is obstructed.

The pressure immediately increases to the peak values dynamic pressure or their reflected pressure and the side on overpressure at any corner of the surface is encountered by the shock wave. The peak values depend upon the surface distance from the source, the size of the explosion, and other factors such as temperature in air and ambient pressure.

The characteristics of the blast wave are described by the overpressure, p_s, versus time, t, curve, the peak initial overpressure, p_{so}, the maximum dynamic pressure, q_o, the duration of positive phase, t_d, and the dynamic pressure, q, versus time, t, curve.

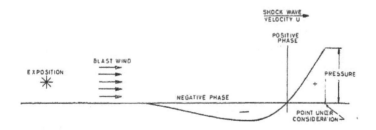

FIGURE 12.17: Shock wave produced by blast (IS 4991:1968)

The intensity of the positive peak rapidly drops to zero. The peak positive overpressure is greater than the maximum negative overpressure, its limiting value being one atmosphere; the duration of the negative phase is 2 to 5 times as long as that of the positive phase. But the overall duration of the positive phase is a few milliseconds.

12.5.2 LOADING EFFECTS DUE TO BLASTS

The phenomena of a blast due to burst above or below the ground surface are difficult. When explosion of a blast happens, a shock front which is circular is propagated away from the burst. The distribution of overpressure at an instant of time along a radial line is shown in Figure 12.18. The shock front moves with a velocity, U, and a peak pressure (p_{so}) which decays behind the shock as indicated. When a shock front strikes a front face of the building, the phenomenon called diffraction produces forces which cause higher pressure due to reflection of the wave from the front face. It is also due to time lag before the overpressure acts on the rare face. The air behind the shock front will also be moving outward at high velocity and it causes drag forces on the building. Thus, the total effect consists of initial diffraction effect, general overpressure (p_s) and the drag.

The dynamic pressure, P_d, is equal to $1/2\ \rho v^2$, where ρ is the density of air and v is the air velocity. The drag pressure is dynamic x drag coefficient C_d. The negative overpressure or suction is generally neglected in design. Only the positive overpressure as a triangular pulse is considered in the design. The curved part of the positive face is approximated by a straight line for computational convenience. Diffraction and drag effects are absent for below ground structures.

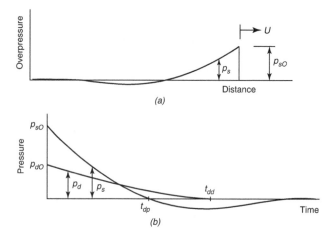

FIGURE 12.18: Pressure-pulse shapes

12.5.3 CRITERIA FOR BLAST RESISTANT DESIGN OF STRUCTURES FOR EXPLOSIONS ABOVE GROUND

Example: Design a RCC building with the following details to resist blast yield of 200 kg at a standout distance of 20 m. The whole structure consists of one part above the ground and one part below the ground (IS 4991:1968).

Single-storey building:
Allowable bearing capacity (qa) = 200 kN/m^2.
Safety factor (sf) = 3
Sub-grade modulus = 40*sf* qa
= 40*3*200
= 24000 kN/m^3
Height of a building = 3 m
Length of a building = 5 m
Breadth of a building = 4 m
Size of beam = 200*300
Size of column = 300*300
Software used is SAFE

After the loadings are computed using IS 4991-1968, the building is analyzed using ETABS and SAFE software (Table 12.15). The loading on the buried structure is computed on an imaginary line away from the building in between the building and the source of the blast, as if it is a vertical surface of any building. This technique will enable one to obtain the effect of soil between the building and the imaginary line in attenuating the effect of the blast.

Here we considered imaginary line away from structure; soil pressure is applied around it and an overpressure is applied to the side of a structure.

A New technique to compute pressures on underground structures. The technique has been developed by the author while guiding a ME project, in Oxford college of Engineering, Bangalore during 2018-19. The technique is enormously useful in computing overpressures on vertical surfaces of underground structures. The problem we come across in computing overpressure on underground structure when there is a blast above the ground surface. The graphs in standard books and codes give overpressures on structures based on the distance from the source of explosion measured horizontally above the ground surface. Now when the structure is below we need to consider the effect of attenuation or dissipation of energy and thus overpresssures due to the presence of the soil. If we model the soil preferably by finite element and consider the soil properties and also the distance of the source it is possible to consider the effect of the soil. For that we need to model the movement of pressure waves through the soil from the source. Therefore if a fictitious vertical surface is imagined in front of the source below the ground surface and obtain the overpressures on it, it is possible to consider the effect of movement of the pressure waves through the soil. Such a technique has been applied to the following problem. It is found effective and gives attenuation of presssures due to soil. Infact, the example on Blast resistant design is added only highlight the techinique as the author feels it seems to be a new technique. A peer reviewed publication is yet to be published on the idea. But there is one small drawback in the method. It does not capture the effect of soil between the source and the fictitious surface however small it is. It is better to consider the surface as close as possible to the source which means that much more volume of soil has to be modelled which takes more CPU time. Except for highlighting the technique the rest of the calculations are the usual stuff available elswhere in books on blast resustant design.

Solution

200kg yield at 20m standoff distance.
x= actual distance/w1/3
x=20/ (0.2)1/3
x=34.199 m
From IS 4991-1968
P_{so}=1.12006 kg/cm2
P_{ro}=3.17 kg/cm2
q_o=0.388kg/cm2
Scaled time t_o and t_d
t_o =25.65*(0.2)1/3=15.002
t_d =16.9614*(0.2)1/3=9.919
M=1.396
a =344 m/s, u = 480.224 = 0.48 m/millisecond

Pressure on building

H=3 m L=5 m B=4 m
S=H or B/2 whichever is less
t_c = 3S/u = 3*2/0.4802=12.49 millisecond
t_t =L/u=5/0.4802=10.41 millisecond
t_r = 4s/u=4*2/0.4802=16.65 millisecond
$t_r > t_d$ no pressure on back face and is zero
For roof and sides C_d = -0.4
P_{so}+ c_d q_o= 1.120+ (-0.4) * 0.388 = 0.964 kg/cm2
Conversion from kg/cm2 to kN/m2
P_{so}+ c_d q_o= 94.56 kN/m2
3.17 kg/cm2=3.17*9.81 N/cm2
= (31.09 N)/ (10-4 m2)
=310.9 kN/m2
Pressure diagram

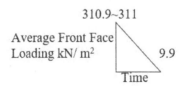

310.9~311

Average Front Face
Loading kN/ m²   9.9

Time

Miscellaneous Aspects 281

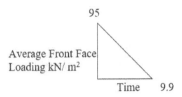

P_{so} = Peak side-on overpressure
P_{ro} = Peak reflected overpressure
M = Mach number for incident shock front

Table 12.15 Results of the structure above and below ground level

Structure	Blast load (Kg)	Standoff distance (m)	Overpressure (kN/m²)	Displacement (mm)	Moment (kN-m)	Max stress (N/mm²)	Shear force (kN)
Below ground level	200	20	311	10.35	1453.79	211.57	8599.16
Above ground level	200	20	311	21.09	401.18	1856.141	9436.20

12.6 RESPONSE OF RCC ASYMMETRIC BUILDINGS SUBJECTED TO EARTHQUAKE GROUND MOTIONS

The work reported here is a part of the dissertation work of student student Davda Karan Kishorebhai. Generally, all buildings will be asymmetrical in plan either due to unsymmetry in mass or stiffness distribution. Unsymmetry in the mass distribution is generally not very common because mostly the major load contribution in all RC buildings is dead load. Dead load will be more or less uniform because slab thickness will be more or less uniform, particularly in shear buildings where we assume the mass is lumped at the floor levels. The major contribution is loads on and from slab.

In view of the above, the major cause for unsymmetry in plan is due to unsymmetry in stiffness; stiffness being the lateral stiffness of the columns, which we assume as $12 \, EI/h^3$. It is easy to find the center of stiffness or rigidity by knowing the lateral stiffness of all the columns. Center of mass can be easily computed; the distance between center of mass and center of rigidity along X-dir. is denoted as e_x and along Y-dir, e_y. Buildings, where either e_x or e_y, or both, is present are called buildings that are unsymmetrical in plan. Such buildings will rotate in plan about vertical axis through the mass center even when the building is subjected to translational ground motions. IS 1893 has very stringent specification regarding such buildings.

Images below show the failure of structures due to irregularity during an earthquake.

12.6.1 STRUCTURAL MODELLING

The different types of models of a framed structure are as follows:
(1) Three-dimensional model: A 3D model, is used when the structure has irregular geometric configuration, torsion due to eccentric distribution in mass and stiffness, etc.

(2) Two-dimensional model: A 2D model, as shown in Fig. 12.19, is used when the structure is symmetric in both principal directions and less torsional responses are expected. It is assumed that behavior will be identical in both directions because of symmetry.

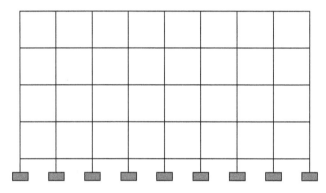

FIGURE 12.19: 2D model

(3) Lumped-mass model: Fig. 12.20 shows a lumped mass model and is simple to analyze and requires less computational efforts when compared to 2D frame models. It was used in the early days to design multi-storey buildings. In such models, as shown in Fig. 12.21, the mass of the building is lumped at each floor level with SDOF in the direction of the displacement and thus DOF in a frame is equal to number of floors.

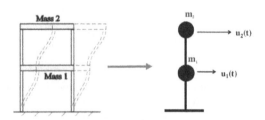

FIGURE 12.20: Lumped-mass model

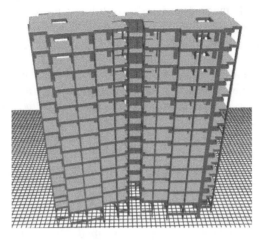

FIGURE 12.21: Elevation

12.6.2 MODELLING AND ANALYSIS OF STRUCTURAL IRREGULARITIES

(1) Structure Details

An analysis is carried out on existing G+12 RCC asymmetric structure for its irregularities defined as per IS 1893 (Part 1): 2002 and 2016 in a structure modeling software ETABS 2017. The plan and elevation are described in Fig. 12.21, respectively. The base dimension in X-dir. is 36.5 m and in Y-dir. is 17.34 m. The structure has two parts shown in plan and among which the second part is at 13° to X-axis because of the site configuration. The analysis was done for the Wing D, and it is clearly seen from the site plan that it needs to be rotated by some angle because of the site configuration.

Structure type: Ordinary Moment Resisting Frame (OMRF)
Total floor: G+12.
Seismic zone: III
Size of columns (mm): 200×700, 200×600.
Size of beam (mm): 200×800, 200×500, 200×450.
Depth of slab (mm): 125, 150.
Response reduction factor: 3.

The analysis was carried out on four structure models described as follows:

Model 1: A model of the building with a floor height of 3m was used as a reference model to differentiate the outcomes for the irregularities.

Model 2: The building is divided into two parts by providing a separation joint between two wings. In the analysis inorder to still keep the two parts of the structure to be connected there is a facility in ETABS to provide what is called a "Gap element" which facilatates the required connection. That is being employed.

Model 3: A ground soft storey model was created to know the behavior of the building for the stiffness irregularity, in which ground storey height was kept as 4 m and without infills, as such buildings are created and used for the parking place due to lack of space available for it.

Model 4: A fourth model was created by adding a basement to the building, which is open storey used for parking, storage, etc. The basement is added in addition to the normal ground storey.

(2) Time-History Data
The building is analyzed using time history method. The accelerogram of previous earthquakes, i.e., El Centro (1940), USA and Bhuj (2001), and India, were applied to the building (Figs. 12.22 and 12.23).

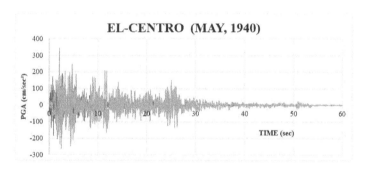

FIGURE 12.22: El Centro (1940) time history

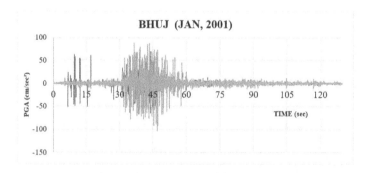

FIGURE 12.23: Bhuj (2001) time history

Miscellaneous Aspects

The details of El Centro (1940) earthquake are as follows:

Magnitude- 6.9.
Duration- 54 s.
Peak ground acceleration- 0.347g at 2.41 s.

(3) The details of Bhuj (2001) earthquake are as follows:
Magnitude- 7.
Duration- 135 s but considered upto 60 s as it reached PGA.
Peak ground acceleration- 0.105 g at 46.94 s.

The analysis results of the building using static and dynamic methods are as follows:

(A) Modal Time Period
Table 12.16 shows the time period of different models in different modes. For the analysis, the total number of modes should be in a manner that summation of modal mass must not be less than 90% of seismic mass. In the table, model 1 and model 2 have the same time period in all the modes because both models are the same, but in

Table 12.16 Modal time period (sec)

MODE NUMBER	TIME PERIOD (sec)			
	MODEL 1	MODEL 2	MODEL 3	MODEL 4
1	2.57	2.57	2.79	3.06
2	2.10	2.10	2.22	2.42
3	1.72	1.73	1.82	1.98
4	0.80	0.80	0.87	0.94
5	0.57	0.57	0.61	0.66
6	0.56	0.56	0.59	0.62
7	0.44	0.44	0.47	0.50
8	0.36	0.36	0.38	0.39
9	0.30	0.30	0.31	0.33
10	0.27	0.27	0.29	0.31
11	0.27	0.27	0.28	0.29
12	0.22	0.22	0.23	0.24
13	0.21	0.21	0.21	0.22
14	0.17	0.17	0.18	0.19
15	0.16	0.16	0.17	0.18
16	0.16	0.16	0.17	0.18
17	0.14	0.14	0.14	0.15
18	0.13	0.13	0.13	0.14
19	0.12	0.12	0.12	0.13
20	0.11	0.11	0.12	0.12

model 2 a gap element in place of separation joint is provided in the model so that the software considers both the wings of the structure as a single structure. The time period of the models 3 and 4 are higher compared to models 1 and 2 as the stiffness on the ground floor is less because of the soft storey.

(B) Modal Mass Participating Factors

Tables 12.17-12.20 below show the mass participating factors of different models in different modes in X and Y dir. as well as rotation about vertical axis in Z dir. From the table, it is seen that the first mode of the structure is influenced by the rotation and it can be concluded that the structure has torsion irregularity.

(C) Base Shear Comparison

Figs. 12.24-12.28 show the base shear in X-dir. and Y-dir. for IS 1893:2002 and 2016. It was seen that base shear in model 1 is high compared to all in each case. The base shear obtained from the linear dynamic method was less than linear static and hence it was scaled to (EQ/RS).

(D) Maximum Top Storey Displacement and Storey drifts

IS 456 and IS 1893 have specified limit for lateral sway of any structure which should not exceed (height/250). Maximum Permissible Lateral Sway in different models as per IS 1893:2002 and 2016.

Table 12.17 Model 1 mass participating factors

MODEL 1			
Mode	Period	Sum X	Sum Y	Sum Z
	sec	%	%	%
1	2.574	17.92	0.17	62.65
2	2.097	18.37	74.01	62.68
3	1.724	77.53	74.4	79.56
4	0.8	79.96	74.41	88.03
5	0.574	81.63	85.9	88.62
6	0.555	92.24	88.52	91.62
7	0.441	93.05	88.52	94.35
8	0.358	95.48	88.54	95.11
9	0.296	95.99	88.57	96.32
10	0.269	96	93.75	96.36
11	0.267	97.07	93.77	96.87
12	0.217	97.58	93.77	97.24
13	0.207	97.79	93.77	97.82
14	0.172	98.26	93.77	97.83
15	0.162	98.27	95.12	98.15
16	0.161	98.27	96.39	98.4
17	0.141	98.54	96.39	98.4
18	0.129	98.54	96.39	98.75
19	0.119	98.72	96.39	98.75
20	0.111	98.72	97.74	98.75

Miscellaneous Aspects

Table 12.18 Model 2 mass participating factors

\multicolumn{5}{c}{MODEL 2}
Mode	Period	Sum X	Sum Y	Sum Z
	sec	%	%	%
1	2.575	18.02	0.19	62.52
2	2.096	18.46	74.05	62.56
3	1.726	77.49	74.41	79.56
4	0.8	79.94	74.42	88.02
5	0.574	81.8	85.66	88.67
6	0.555	92.22	88.52	91.62
7	0.441	93.05	88.52	94.35
8	0.358	95.48	88.54	95.11
9	0.296	95.98	88.57	96.32
10	0.269	96	93.74	96.36
11	0.267	97.06	93.77	96.87
12	0.217	97.58	93.77	97.23
13	0.207	97.79	93.77	97.82
14	0.172	98.26	93.77	97.83
15	0.162	98.27	95.1	98.15
16	0.161	98.27	96.39	98.4
17	0.141	98.54	96.39	98.4
18	0.129	98.54	96.39	98.75
19	0.119	98.72	96.39	98.75
20	0.111	98.72	97.74	98.75

Table 12.19 Model 3 mass participating factors

\multicolumn{5}{c}{MODEL 3}
Mode	Period	Sum X	Sum Y	Sum Z
	sec	%	%	%
1	2.791	19.3	0.12	65.5
2	2.221	19.67	76.66	65.52
3	1.825	80.85	77.03	83.32
4	0.866	83.04	77.04	90.92
5	0.61	84.31	88.73	91.37
6	0.587	95.01	90.87	94.5
7	0.471	95.51	90.88	96.42
8	0.376	97.2	90.9	96.99
9	0.312	97.43	90.93	97.66
10	0.285	97.44	95.43	97.7
11	0.277	97.98	95.43	97.97
12	0.226	98.15	95.43	98.17
13	0.214	98.24	95.44	98.39
14	0.177	98.4	95.44	98.4
15	0.17	98.4	97.34	98.41
16	0.167	98.41	97.37	98.62
17	0.144	98.49	97.37	98.62
18	0.132	98.49	97.37	98.74
19	0.121	98.54	97.37	98.74
20	0.115	98.54	98.21	98.74

Table 12.20 Model 4 mass participating factors

Mode	Period	Sum X	Sum Y	Sum Z
	sec	%	%	%
1	3.055	19.84	0.12	66.96
2	2.424	20.29	78.32	66.97
3	1.981	82.37	78.77	85.2
4	0.937	84.32	78.78	92.08
5	0.664	84.58	91.33	92.19
6	0.624	95.64	92.01	95.48
7	0.502	96.02	92.02	97.02
8	0.391	97.56	92.04	97.52
9	0.331	97.75	92.08	98.06
10	0.31	97.76	95.95	98.1
11	0.288	98.27	95.95	98.32
12	0.239	98.42	95.96	98.53
13	0.224	98.56	95.96	98.71
14	0.186	98.76	95.98	98.73
15	0.183	98.77	97.57	98.73
16	0.176	98.79	97.57	98.98
17	0.153	98.95	97.57	98.98
18	0.141	98.95	97.57	99.17
19	0.13	99.09	97.57	99.17
20	0.123	99.09	98.32	99.17

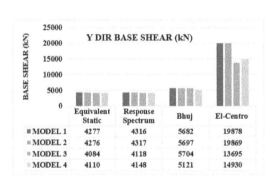

FIGURE 12.24: Y-dir. base shear (IS1893:2016)

The graphs in Table 12.21 and Figs. 12.29-12.32 with the displacement values show that model 1 has less lateral sway and is also in the limiting range specified by the standards. Though in the results of top displacement it is seen that there is less difference in top displacement, but when the storey drift results (Figs. 12.32 and 12.33) are seen, it is clear that in the drift in-ground storey is higher because of soft storey effect.

Miscellaneous Aspects

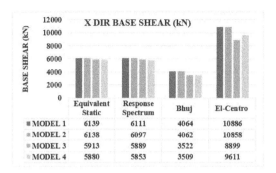

FIGURE 12.25: X-dir. base shear (IS1893:2016)

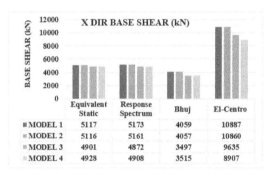

FIGURE 12.26: X-dir. base shear (IS1893:2002)

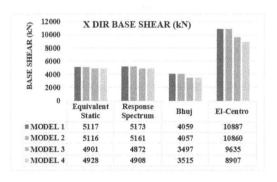

FIGURE 12.27: Y-dir. base shear (IS1893:2002)

Figs. 12.32 and 12.33 show the storey drift along X-dir. and Y-dir., respectively. Model 1 and model 2 have the same storey drifts while model 3 and model 4 in the ground storey, the storey drifts are higher because of the soft storey effect.

Table 12.21 Permissible lateral sway

Model	Lateral Sway (mm)
MODEL 1	163
MODEL 2	163
MODEL 3	168
MODEL 4	179

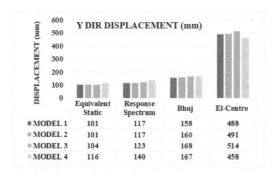

X DIR DISPLACEMENT (mm)	Equivalent Static	Response Spectrum	Bhuj	El-Centro
MODEL 1	139	88	96	261
MODEL 2	139	88	97	261
MODEL 3	147	91	106	228
MODEL 4	165	104	105	249

FIGURE 12.28: Top storey displacement X-dir. (IS1893:2002)

Y DIR DISPLACEMENT (mm)	Equivalent Static	Response Spectrum	Bhuj	El-Centro
MODEL 1	101	117	158	488
MODEL 2	101	117	160	491
MODEL 3	104	123	168	514
MODEL 4	116	140	167	458

FIGURE 12.29: Top storey displacement Y-dir. (IS1893:2002)

(E) Torsional Irregularity

Torsional irregularity is one among many factors which can damage the building severely. The irregularity depends on several factors such as geometry of plan, the arrangement of structural components and their dimensions and also on the storey numbers. The irregularity ratio governs the multi-directional behavior of the building and also recognizes the ability of elements resisting seismic impact. Torsional irregularity is taken into account when max. storey drift, along with design eccentricity, is calculated. As per the IS1893:2002, torsional irregularity in a structure exists when the displacement ratio ($\Delta_{max}/\Delta avg$) of the storey at two ends exceeds 1.2. Seismic code IS1893:2016 says it exists when the displacement ratio ($\Delta_{max}/\Delta min$) at the two ends of the structure exceeds 1.5. The limiting ratio as per IS 1893:2002 ($\Delta_{max}/\Delta avg$)

Miscellaneous Aspects

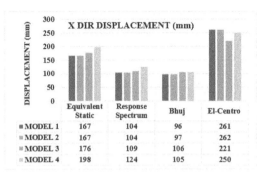

FIGURE 12.30: Top storey displacement X-dir. (IS1893:2016)

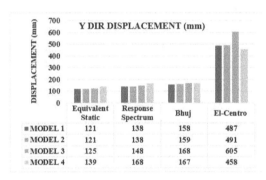

FIGURE 12.31: Top storey displacement Y-dir. (IS1893:2016)

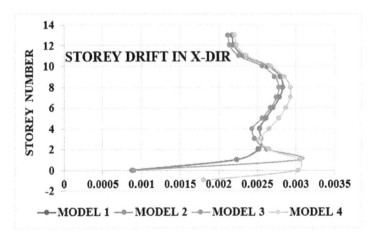

FIGURE 12.32: Storey drift in X-dir.

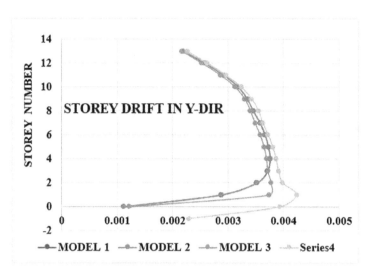

FIGURE 12.33: Storey drift in Y-dir.

≤ 1.2 defined, and as per IS 1893:2016 ($\Delta_{max}/\Delta_{min}$) should be in the range of 1.5-2.0 and if the ratio exceeds 2.0, the structure configuration in that direction needs to be revised. When the building is subjected to a torsional irregularity, it has differential deformation in plan and will influence the performance of the building, and thus the design of resisting elements is done to reduce the torsional effects.

Tables 12.22 and 12.23 show the torsional irregularity ratios ($\Delta_{max}/\Delta avg$) and ($\Delta_{max}/\Delta min$) as defined by IS 1893:2002 and 2016, respectively, of different models for the response spectrum analysis of the building, and was it seen that all models have torsional irregularity because of plan deformation in geometry.

The results show that the irregularity ratio as per both seismic codes exceeds the limiting range in Y-dir. IS1893:2016 clearly mentioned that if the ratio exceeds 2.0, the configuration needs to be revised. In X-dir. ratio is within the limit.

(F) Stiffness Irregularity (Soft Storey)

A soft storey is one which has more openings, such as in one or more floor there are more windows or large openings for doors, which may result in a decrease in stiffness of the storey. A typical soft storey structure is defined as a structure having three or more storeys in which ground level is kept open for parking or the stores having large openings for doors or windows or the structure designed for less strength on the upper storey also known as top soft storey. The performance of such structures is less during an earthquake and in the past earthquakes, most of the buildings got damaged at soft storey termed as soft storey failure.

Miscellaneous Aspects

Table 12.22 Torsional irregularity ratios for different models ($\Delta_{max}/\Delta_{avg}$)
(IS 1893:2002)

STORY	MODEL 1 X	MODEL 1 Y	MODEL 2 X	MODEL 2 Y	MODEL 3 X	MODEL 3 Y	MODEL 4 X	MODEL 4 Y
12	1.12	1.26	1.12	1.26	1.12	1.27	1.12	1.27
11	1.12	1.27	1.12	1.27	1.12	1.28	1.12	1.28
10	1.12	1.28	1.11	1.28	1.11	1.29	1.11	1.29
9	1.11	1.29	1.11	1.29	1.11	1.30	1.11	1.30
8	1.11	1.30	1.11	1.30	1.10	1.31	1.10	1.30
7	1.10	1.31	1.10	1.31	1.10	1.31	1.10	1.31
6	1.10	1.32	1.10	1.32	1.10	1.33	1.10	1.32
5	1.09	1.33	1.09	1.33	1.11	1.34	1.11	1.33
4	1.10	1.34	1.10	1.35	1.12	1.35	1.12	1.35
3	1.10	1.36	1.10	1.36	1.13	1.38	1.13	1.37
2	1.10	1.39	1.10	1.39	1.13	1.40	1.14	1.39
1	1.10	1.42	1.10	1.42	1.13	1.44	1.14	1.42
0	1.09	1.46	1.09	1.46	1.12	1.49	1.15	1.46
BASEMENT							1.14	1.52

Table 12.23 Torsional irregularity ratios for different models ($\Delta_{max}/\Delta_{min}$)
(IS 1893:2016)

STORY	MODEL 1 X	MODEL 1 Y	MODEL 2 X	MODEL 2 Y	MODEL 3 X	MODEL 3 Y	MODEL 4 X	MODEL 4 Y
12	1.27	1.71	1.27	1.71	1.27	1.73	1.27	1.73
11	1.27	1.74	1.27	1.75	1.26	1.77	1.27	1.77
10	1.26	1.78	1.26	1.79	1.25	1.80	1.26	1.80
9	1.25	1.82	1.25	1.82	1.24	1.84	1.24	1.84
8	1.24	1.85	1.23	1.85	1.23	1.88	1.23	1.87
7	1.22	1.89	1.22	1.89	1.22	1.92	1.22	1.90
6	1.21	1.93	1.21	1.93	1.22	1.96	1.22	1.95
5	1.21	1.98	1.21	1.99	1.25	2.02	1.25	2.00
4	1.22	2.05	1.22	2.05	1.27	2.09	1.27	2.06
3	1.22	2.14	1.23	2.14	1.28	2.20	1.30	2.15
2	1.22	2.27	1.23	2.27	1.30	2.35	1.32	2.27
1	1.21	2.45	1.21	2.45	1.29	2.59	1.34	2.45
0	1.19	2.67	1.19	2.67	1.27	2.92	1.34	2.73
BASEMENT							1.33	3.19

A soft storey has stiffness 70% of the storey above or 80% of the average of above three storeys defined as per code IS1893:2002 and 80% less than the storey above when RC walls are provided in the building as per IS1893:2016.

Tables 12.24 and 12.25 show the value of stiffness of the different models. It is shown that in soft storey model 3, the stiffness of the ground level in X-dir. is less

Table 12.24 Building stiffness along Y-dir

STORY	MODEL 1 kN/m	MODEL 2 kN/m	MODEL 3 kN/m	MODEL 4 kN/m
12	135833	135974	125801	114903
11	220534	220674	208398	193343
10	255017	255129	244134	229338
9	269124	269182	259123	245584
8	275581	275571	266237	253884
7	280245	280296	271756	259599
6	286708	286680	279041	266652
5	300685	300591	289945	274850
4	324118	323946	308401	287515
3	357438	357181	335010	306483
2	402838	402449	369315	331593
1	473573	472980	409046	360644
0	640405	639461	318363	320675
Basement				657699

Table 12.25 Building stiffness along X-dir

STORY	MODEL 1 kN/m	MODEL 2 kN/m	MODEL 3 kN/m	MODEL 4 kN/m
12	167130	166442	151719	143535
11	306186	304859	286778	272273
10	382326	380600	366030	349450
9	407757	405985	394484	379005
8	427820	425985	415674	399925
7	448249	446352	435340	417518
6	474653	472690	458822	436519
5	512410	510390	491891	461770
4	563914	561852	539512	498178
3	626907	624883	604567	550011
2	708355	705745	659653	615853
1	828136	825353	712676	653751
0	1065720	1062753	519295	562048
Basement				658200

than 70% than first level. While the stiffness of the ground level along Y-dir. is not less than 70%. Thus, the structure is irregular in stiffness along X-dir.

In this study, an analysis of existing G+12 RCC structure done for irregularities defined as per IS 1893:2002 and IS 1893:2016. The examination for the lateral load is done using the time history of the past earthquake, i.e., El Centro (May 1940) and Bhuj (Jan. 2001) on four different models, and the conclusions are listed below:

the analysis showed that base shear in model 1 is the largest. The lateral sway is more than the allowable limit in all the models. All the models undergo torsional, irregularity.

The structural configuration of model 2, model 3, and model 4 should be revised as torsional irregularity ratio exceeds 2.0 as per IS1893:2016. The soft-storey model 3 has stiffness irregularity, which may result in soft storey failure.

A separation joint anywhere in the structure could be provided to make it less asymmetric.

Irregularity in the structure either in mass or stiffness will increase the forces to be resisted by the structural elements which will affect the overall cost of the structure.

When the results compared as per both the codes, it showed that the latest edition of the seismic law is stringent towards the structural irregularity.

Due to increase in the limits of the irregularity, as per IS1893:2016, serious consideration to the code restriction should be given.

Concluding Remarks. Although the results reported above from the various dissertation theses appear to be sketchy the author believes and hopes that they open-up some new topics for further research which will be particularly useful to the earthquake and blast resistant analysis and design.

References and Suggestions for Further Reading

References Chapters 1–7

1. Elements of Vibration Analysis by Leonard Meirovitch, McGraw-Hill Book Company, 1986.
2. Introduction to Structural Dynamics by J.M. Biggs, McGraw-Hill Book Company, 1964.
3. Structural Dynamics, Theory and Computation by Mario Paz, CBS Publishers and Distributors, 2004.
4. Dynamics of Structures, Theory and Applications to Earthquake engineering by Anil K. Chopra, Pearson Education Asia, 2001.
5. Dynamics of Structures by Ray W. Clough, Joseph Penzien, McGraw-Hill Book Company, 1982.
6. Mechanical Vibrations by A.H. Church, John Wiley and Sons, Inc., New York and London, 1957.
7. Shock, Impact and Explosion, Structural Analysis and Design by M.Y.H. Bangash, Springer, 2008.
8. Vibration, Fundamentals and Practice by Clarence W. de Silva, CRC Press, 2005.
9. Theory of Vibration with Applications by W.T. Thomson, CBS Publishers and Distributors, 1972.
10. Mechanical Vibrations by J.P. Den Hartog, Dover Publications, 1985.
11. Structural Dynamics, An Introduction to Computer Methods by Roy R. Craig, Jr., John Wiley and Sons, Inc., 1981.
12. Iyengar, R. N., Elements of Mechanical Vibration. International Publishing House, (Pvt.) Ltd., New Delhi, Bangalore.
13. Freeman, S. A., Nicoletti, J. P. and Tyrell, J.V., Evaluations Of Existing Buildings For Seismic Risk - A Case Study Of Puget Sound Naval Shipyard, Bremerton, Washington Proceedings of the 1st U.S. National Conference on Earthquake Engineering, EERI, Berkeley, CA, 1975, 113-122.
14. Freeman, S.A., Review of the Development of the Capacity Spectrum Method. ISET Journal of Earthquake Technology, pp No. 438, Vol. 41, No.1, March 2004, pp. 1-13.
15. Fajfar, P. and Fischinger, M., N2- A Method For Non-Linear Seismic Analysis Of Regular Buildings, Proceedings of the 9th World Conference on Earthquake Engineering, Tokyo: Maruzen, 1989, Vol.5:111-116.
16. Raghu Prasad, B.K., Seethha Ramaiah A. and Singh, A.K., Capacity Spectrum for structures asymmetric in the plan, 13th World Conference on Earthquake Engineering, Vancouver, B.C., Canada, August 1-6, 2004, pp. 2653.
17. Fajfar, P., Structural Analysis in Earthquake Engineering- A breakthrough of Simplified Non-Linear Methods. Elsevier Publication. 12th European Conference on Earthquake Engineering, pp. 843.

References Chapters 8–10

1. Housner, G.W. "Limit Design of Structures to Resist Earthquakes", Proceedings of the First World Conference on Earthquake Engineering, Berkeley, California, 1956, pp. 5-1 to 5-13.
2. Tanabashi, R. "Studies on Nonlinear Vibrations of Structures Subjected to Destructive Earthquakes", Proceedings of the First World Conference on Earthquake Engineering, Berkeley, California, 1956, pp. 6-1 to 6-13.
3. Bycroft, G.N., Murphy, M.J. and Brown, K.J, "Electric Analog for Earthquake Yield Spectra", Journal of the Engineering Mechanics Division, Proc. of the ASCE, vol. 85, No. EM4, Oct 1959, pp. 43-64.
4. Blume, J.A., "Structural Dynamics in Earthquake Resistant Design", Transactions ASCE, 125, 1960, pp. 1088-1139.
5. Clough, R.W., "Earthquake Response of Structures", Earthquake Engineering, Prentice-Hall, Inc., N.J., 1970, pp. 307-334.
6. Pan, S.L., "Influence of ductility on the response of simple structures to earthquake motions", Ph.D. Dissertation, Univ. of Illinois, Dept. of Civil Engineering, 1951.
7. Housner, G.W., "Behaviour of structures during Earthquakes", Journal of the Engineering Mechanics Division, Proc. Of the ASCE, vol. 85, No. EM4, Oct 1959, pp. 109-129.
8. Housner, G.W., " The Plastic Failure of Frames During Earthquakes", Proceedings of the Second World Conference on Earthquake Engineering vol. II, Tokyo, Japan, 1960, pp. 997 to 1012.
9. Veletsos, A.S. and Newmark, N.M. "Effect of the Inelastic Behaviour on the Response of Simple Systems to Earthquake Motions", Proceedings of the Second World Conference on Earthquake Engineering vol. II, Tokyo, Japan, 1960, pp. 895 to 912.
10. Penzien, J. "Dynamic Response of Elasto-plastic Frames", Journal of Structural Division, Proc. Of ASCE, vol. 86, No. ST7, July 1960, pp. 81-94.
11. Odaka, T. and Horie, F. "A Study on the Optimum value of a Seismic Coefficient", Proceedings of the Third World Conference on Earthquake Engineering vol. II, Auckland and Wellington, New Zealand, 1965, pp. II-399 to II-420.
12. Thomaides, S.S. "Earthquake Response of the Systems with Bilinear Hysteresis", Journal of Structural Division, Proc. Of ASCE, vol. 90, No. ST4, August 1964, pp. 123-143.
13. Berg, G.V. and Thomaides, S.S. "Energy Consumption by Structures in Strong-Motion Earthquake", Proceedings of the Second World Conference on Earthquake Engineering vol. II, Tokyo, Japan, 1960, pp. 681 to 697.
14. Pereira, J.M.J. " Behaviour of an Elasto-Plastic Oscillator acted by Random Vibration", Proceedings of the Third World Conference on Earthquake Engineering vol. II, Auckland and Wellington, New Zealand, 1965, pp. II-491 to II-501.
15. Jennings, P.C., "Response of Yielding Structures to Statistically Generated Ground Motion", Proceedings of the Third World Conference on Earthquake Engineering vol. II, Auckland and Wellington, New Zealand, 1965, pp. II-237 to II-257.
16. Jennings, P.C., "Periodic Response of a General Yielding Structure", Journal of the Engineering Mechanics Division, Proc. Of the ASCE, vol. 90, No. EM2, April 1964, pp. 132-166.
17. Hudson, D.E., "Equivalent Viscous Friction for Hysteretic Systems with Earthquake-Like Excitations", Proceedings of the Third World Conference on Earthquake Engineering vol. II, Auckland and Wellington, New Zealand, 1965, pp. II-185 to II-201.
18. Ruge, A.C., "The determination of Earthquake Stresses in Elastic Structures by means of Models", Bulletin of the seismological society of America, vol. 24, No.3, July 1934.

References and Suggestions for Further Reading 299

19. Jacobsen, L.S., "Dynamic Behavior of Simplified Structures up to point of Collapse", Proceedings of the Symposium on Earthquake and Blast Effects on Structures, Los Angeles, California, 1952.
20. Husid, R. "Gravity Effects on the Earthquake Response of Yielding Structures", Ph.D. Thesis, California Institute of Technology, 1967.
21. Jennings, P.C. and Husid, R. "Collapse of Yielding Structures During Earthquakes", Journal of the Engineering Mechanics Division, Proc. of the ASCE, vol. 94, No. EM5, October 1968, pp. 1045-1065.
22. Liu, S.C. "Earthquake Response Statistics of Non-linear Systems", Journal of the Engineering Mechanics Division, Proc. of the ASCE, vol. 95, No. EM2, April 1969, pp. 397-419.
23. Sun, C.K., Berg, G.V. and Hanson, R.D. "Gravity Effect on Single-Degree Inelastic System", Journal of the Engineering Mechanics Division, Proc. of the ASCE, vol. 99, No. EM1, February 1973, pp. 183-200.
24. Tani, S. and Soda, S. "Vertical Load Effects on Structural Dynamics", Proceedings of the Sixth World Conference on Earthquake Engineering, vol. III, New Delhi, India, 1977, pp. 3-55 to 3-60.
25. Iwan, W.D. "A Model for the Dynamic Analysis of Deteriorating Structures", Proceedings of the Fifth World Conference on Earthquake Engineering, Rome, 1973, p. 222.
26. Veletsos, A.S. "Maximum Deformations of Certain Nonlinear Systems", Proceedings of the Fourth World Conference on Earthquake Engineering, Casilla 2777, Santiago, Chile, vol. 2, 1969, pp. 155-170.
27. Clough, R.W. "Dynamic Effects of Earthquake", Journal of Structural Division, Proc. Of ASCE, vol. 86, No. ST4, April 1960, pp. 49-65.
28. Berg, G.V. "Response of Multi-Story Structures to Earthquakes", Journal of the Engineering Mechanics Division, Proc. of the ASCE, vol. 87, No. EM2, April 1961, pp. 1-16.
29. Goel, S.C. and Berg, G.V. "Inelastic Earthquake Response of Tall Steel Frames", Journal of Structural Division, Proc. Of ASCE, vol. 94, No. ST8, August 1968, pp. 1907-1934.
30. Clough, R.W., Benuska, K.L. and Vilson, R.L. "Inelastic Earthquake Response of Tall Buildings", Proceedings of the Third World Conference on Earthquake Engineering, vol. II, Auckland and Wellington, New Zealand, 1965, pp. II-185 to II-201.
31. Clough, R.W. and Benuska, K.L. "Nonlinear Earthquake Behaviour of Tall Buildings", Journal of the Engineering Mechanics Division, Proc. of the ASCE, vol. 93, No. EM3, June 1967, pp. 129-146.
32. Goel, S.C. "P-Δ and Axial Column Deformation in Aseismic Frames", Journal of Structural Division, Proc. of ASCE, vol. 95, No. ST8, August 1969, pp. 1693-1711.
33. Anderson, J.C. and Bertero, V.V. "Effects of Gravity Loads and Vertical Acceleration on the Seismic Response of Multi Storey Frames", Proceedings of the Fifth World Conference on Earthquake Engineering, Rome, 1973, p. 369.
34. Mahin, S.A. and Bertero, V.V. "Nonlinear Seismic Response Evaluation - Charaima Building", Journal of Structural Division, Proc. of ASCE, vol. 100, No. ST6, June 1974, pp. 1225-1242.
35. Sun, C.K. "Hybrid Techniques for Inelastic Dynamic Problems", Journal of Structural Division, Proc. of ASCE, vol. 101, No. ST4, April 1975.
36. Tanabashi, R., Nakamura, T. and Ishida. "Gravity Effect on the Catastrophic Dynamic Response of Strain Hardening Multistorey Frames", Proceedings of the Fifth World Conference on Earthquake Engineering, Rome, 1973, p. 268.
37. Weingarten, V.I., Masri, S.P. and Lashkari, M. "Effect of Gravity Loading on Earthquake Response of Cooling Towers", Proceedings of the Fifth World Conference on Earthquake Engineering, Rome, 1973, p. 269.

38. Goel, S.C. "Seismic Behaviour of Multistorey K-Braced Frames Under Combined Horizontal and Vertical Ground Motion", Proceedings of the Sixth World Conference on Earthquake Engineering, vol. III, New Delhi, India, 1977, pp. 3-199 to 3-204.
39. Ruiz, P. and Penzien, J. "Stochastic Seismic Responses of Structures", Journal of the Engineering Mechanics Division, Proc. of the ASCE, vol. 97, No. EM2, June 1971, pp. 441-456.
40. Brijesh, Chandra and Chandrasekaran, A.C. "Elasto-Plastic Behaviour of Multistorey Frames", Proceedings of the Fourth Symposium on Earthquake Engineering, Roorkee, India, November 1970, pp. 266-271.
41. Arya, A.S. "Inelastic and Reserve Energy Analysis of Multistoreyed Buildings", Proceedings of the Fifth World Conference on Earthquake Engineering, Rome, 1973, p. 277.
42. Anderson, J.C. and Gupta, R.P. "Earthquake Resistant Design of Unbraced Frames", Journal of Structural Division, Proc. of ASCE, vol. 98, No. ST11, November 1972, pp. 2523-2539.
43. Veletsos, A.S. and Vann, W.P. "Response of Ground-Excited Elasto-Plastic Sytems", Journal of Structural Division, Proc. of ASCE, vol. 97, No. ST4, April 1971, pp. 1257-1281.
44. Gupta, R.P. and Goel, S.C. "Dynamic Analysis of Staggered Frame Truss Framing System", Journal of Structural Division, Proc. of ASCE, vol. 98, No. ST7, July 1972, pp. 1475-1492.
45. Goel, S.C. and Hanson R.D. "Seismic Behaviour of Multistory Braced Steel Frames", Journal of Structural Division, Proc. of ASCE, vol. 100, No. ST1, January 1974, pp. 79-95.
46. Imbeaul, T., F. A. and Nielsen, N.N. "Effect of Degrading Stiffness on Response of Mulitstory Frames Subjected to Earthquakes", Proceedings of the Fifth World Conference on Earthquake Engineering, Rome, 1973, pp. 220.
47. Iwan, W.L. "The Response of Simple Stiffness Degrading Structures", Proceedings of the Sixth World Conference on Earthquake Engineering, vol. III, New Delhi, India, 1977, pp. 3-121 to 3-126.
48. Otani, S. and Sozen, M.A. "Simulated Earthquake Tests on RC Frames", Journal of Structural Division, Proc. of ASCE, vol. 100, No. ST3, March 1974, pp. 687-701.
49. Crandall, S.H. and Mark, W.D. "Random Vibration in Mechanical Systems", Academic Press, 1963.
50. Curtis, A.J. and Boykin, T.R. (Jr.). "Response of Two-Degree-of-Freedom System to White Noise Base Excitation", Journal of the Acoustical Society of America, vol.33, 1061, pp. 655-663.
51. Vasudev Rao, P. and Jagadish, K.S. "Vibration Absorber Under Earthquake Type Base Excitation", Journal of Institution of Engineers (India), vol.50, No.9, Part C15, May 1970, pp. 271-274.
52. Roberson, R.E. "Synthesis of a Nonlinear Dynamic Vibration Absorber", Journal of Franklin Institute, vol.254, No.3, 1952, pp. 205-220.
53. Pipes, L.A. "Analysis of a Nonlinear Dynamic Vibration Absorber", Journal of Applied Mechanics, Trans. ASME, vol. 20, 1953, pp. 515-518.
54. Arnold, F.r. "Steady-State Behaviour of System Provided With Nonlinear Dynamic Vibration Absorber", Journal of Applied Mechanics, Trans. ASME, vol. 22, 1955, pp. 487-492.
55. Carter, W.J. and Liu, F.C. "Steady-State Behaviour Nonlinear Dynamic Vibration Absorber", Journal of Applied Mechanics, Trans. ASME, vol. 28, 1961, pp. 67-70.
56. Bauer, H.F. "Steady State Harmonic and Combination Response of a Nonlinear Dynamic Vibration Absorber", Journal of Applied Mechanics, Trans. ASME, vol. 33, 1966, pp. 213-216.

References and Suggestions for Further Reading

57. Chandrasekaran, A.R. and Gupta, Y.P. "Vibration Absorber of Single Degree Freedom System", Proceedings of third Symposium on Earthquake Engineering, Roorkee, November 1966, pp. 23-32.
58. Gupta, Y.P. and Chandrasekaran, A.R. "Dry Friction Type Absorber for Earthquake Excitations", Journal of Institution of Engineers (India), vol.48, No.11, Part C16, July 1968, pp. 1654-1663.
59. Wirshing, P.H and Yao, J.T.P. "A Statistical Study of Some Design Concepts in Earthquake Engineering", Technical Report CE-21 (70) NSF-065, Bureau of Engineering Research, The University of New Mexico, Albuquerque, New Mexico, May 1970.
60. Ohno, S., Watari, A. and Sano, I. "Optimum Tuning of the Damper to Control Response of Structures to Earthquake Ground Motion", Proceedings of the Sixth World Conference on Earthquake Engineering, vol. III, New Delhi, India, 1977, pp. 157-181.
61. Den Hartog, J.P. "Mechanical Vibrations", 1956, McGraw-Hill Book Company, Inc., New York.
62. Indian Standard Criteria for Earthquake Resistant Design of Structures, IS 1983-1966.
63. Seismic Code of Structural Engineers Association of California (October 1959).
64. Rosenblueth, E. "The Earthquake of 28th July 1957, in Mexico City", Proceedings of the Second World Conference on Earthquake Engineering vol. I, Tokyo, Japan, 1960, pp. 359-379.
65. Newmark, N.M. "Torsion in Symmetrical Buildings", Proceedings of the Fourth World Conference on Earthquake Engineering vol. II, Santiago, Chile, A-3 pp. 19-32.
66. Housner, G.W. "The Plastic Failure of Frames During Earthquakes", Proceedings of the Second World Conference on Earthquake Engineering vol. II, Tokyo, Japan, 1960, pp. 997-1012.
67. Bustamante, J.T. and Rosenblueth, E. "Building Code Provisions on Torsional Oscillations", Proceedings of the Second World Conference on Earthquake Engineering vol. II, Tokyo, Japan, 1960, pp. 879-894.
68. Ayre, R.S. "Methods for Calculating the Earthquake Response of Shear Buildings", Proceedings of the World Conference on Earthquake Engineering, Berkely, California, 1956, pp. 1-24.
69. Gibson, R.E., Moody, M.L. and Ayre, R.S. "Response Spectrum Solution for Earthquake Analysis of Unsymmetrical Multi-storeyed Buildings", Bulletin of the Seismological Society of America, vol. 62, Feb 1972, pp. 215-229.
70. Skinner, R.I., Skilton, D.W.C and Laws, D.A. "Unbalanced Buildings and Buildings with Light Towers Under Earthquake Forces", Proceedings of the Third World Conference on Earthquake Engineering vol. II, Auckland and Wellington, New Zealand, 1965, pp. 586-602.
71. Medearis, K. "Coupled Bending and Torsional Oscillations of a Modern Skyscraper", Bulletin of the Seismological Society of America, vol. 56, August 1966, pp. 937-946.
72. Douglas, M. and Trabert, T.E. "Coupled Torsional Dynamic Analysis of a Multi-Storeyed Building", Bulletin of the Seismological Society of America, vol. 63, June 1973, pp. 1025-1039.
73. Shepard, R. and Donald, R.A.H. "Seismic Response of Torsionally Unbalanced Buildings", Journal of Sound and Vibration, vol. 6, No.1, 1967, pp. 20-37.
74. Housner, G.W. and Outinen, H. "The Effect of Torsional Oscillations on Earthquake Stresses", Bulletin of the Seismological Society of America, vol. 48, July 1958, pp. 221-229.

75. Shiga, T. "Torsional Vibration of the Multistoreyed Buildings", Proceedings of the Third World Conference on Earthquake Engineering vol. II, Auckland and Wellington, New Zealand, 1965, pp. 569-584.
76. Housner, J.B. "Modal Coupling and Earthquake Response of Tall Buildings", Ph.D. Thesis, California Institute of Technology, Pasadena, California, 1971.
77. Blume, J.A. "Structural Dynamics in Earthquake Resistant Design", Journal of Structural Division, Proc. of ASCE, vol. 84, No. ST4, March 1958, pp. 1-25.
78. Keintzel, E. "On the Seismic Analysis of Unsymmetrical Storeyed Buildings", Proceedings of the Fifth World Conference on Earthquake Engineering, Rome, 1973, pp. 10, Session 1B.
79. Ayre, R.S. "Experimental Response of an Asymmetric, One-storey Building Model to an Idealized Transient Ground Motion", Bulletin of the Seismological Society of America, vol. 33, July 1943, pp. 91-119.
80. Penzien, J. and Chopra, A.K. "Earthquake Response of an Appendage on Multistorey Building", Proceedings of the Third World Conference on Earthquake Engineering vol. II, Auckland and Wellington, New Zealand, 1965, pp. 476-490.
81. Strong-Motion Earthquake Accelerograms - Digitised and Plotted Data, Earthquake Engineering Research Laboratory, California Institute of Technology, Pasadena, California, 1971.
82. Nigam, N.C. "Yielding in Framed Structures Under Dynamic Loads", Journal of the Engineering Mechanics Division, Proc. of the ASCE, vol. 96, No. EM5, Oct 1970, pp. 687-709.
83. Kobori, T., Minai, R. and Fujiwara. "Earthquake Response of Frame Structures Composed of Inelastic Members", Proceedings of the Fifth World Conference on Earthquake Engineering, Rome, 1973, p. 221.
84. Pecknold, D.A. "Inelastic Structural Response to 2D Ground Motion", Journal of the Engineering Mechanics Division, Proc. of the ASCE, vol. 100, No. EM5, Oct 1974, pp. 949-963.
85. Kachanov, L.M. "Foundations of the Theory of Plasticity", North Holland Publishing Co., 1971, Amsterdam, London.
86. Prager, W. "An Introduction to Plasticity", Addison-Wesley Publishing Co., Inc., 1955, USA, England.
87. Massonet, C.E. and Save, M.A. "Plastic Analysis and Design of Plates, Shells and Disks", 1972, North-Holland Publishing Company, Amsterdam, London.
88. Hodge, P.G. Jr. "Plastic Analysis of Structures", 1959, McGraw-Hill Book Co. Inc., New York.
89. Calladine, C.R. "Engineering Plasticity", 1969, Pergamon Press, London, New York.
90. Olszak, W., Mroz, Z. and Perzyna, P. "Recent Trends in the Development of the Theory of Plasticity", 1963, Pergamon Press, London, New York.

References Chapter 12

1. R. Gayathri, B.K. Raghu Prasad, Amarnath, K. Retrofitting Methods for RCC Structures, IJEAT, Vol.8, Issue 05, June 2019.
2. Swati Ricke, B.K. Raghu Prasad, Amarnath, K. Response Spectrum Analysis of Tall Building using PYTHON, IJERT, Vol.8, Issue 07, July 2019.
3. Muniswamy Rajakumar - Proprietor and Founder of Vidhaatri and PMGR Groups.
4. Amir Khan, B.K. Raghu Prasad, Amarnath, K. Assessment of Hybrid Building, IJEAT, Vol.8, Issue 06, August-2019.

References and Suggestions for Further Reading

5. Sharlin Sheeba, B.K. Raghu Prasad, Amarnath, K. Analysis and Design of Blast Resistant Structures, IJERT, Vol.8, Issue 07, July 2019.
6. Davda Karan Kishorbhai, B.K. Raghu Prasad, Amarnath, K. Response of RCC Asymmetric Building Subjected to Earthquake Ground Motions, IJERT, Vol.8, Issue 07, July 2019.
7. B.K. Raghu Prasad and Jagadish, K. S., "Inelastic torsional response of a single-storey framed structure," Journal of engineering mechanics, ASCE, vol.115, No.08, pp. 1782-1797, August 1989.
8. B.K. Raghu Prasad, A. Seethha Ramaiah and A.K. Singh. Capacity Spectrum for Structures Asymmetric in the Plan, 13th WCEE, p. 2653.
9. K. S. Jagadish, B. K. Raghu Prasad and P. Vasudeva Rao, The Inelastic Vibration Absorber Subjected To Earthquake Ground Motions, Earthquake Engineering and Structural Dynamics, Vol 7, 317-326 (1979) John Wiley & Sons, Ltd.
10. B.K. Raghu Prasad, N. Lakshmanan, K. Muthumani, N. Gopalakrishnan and R. Sreekala. Seismic Damage Estimation through Measurable Dynamic Characteristics, Computers, and Concrete, Vol. 4, No. 3, 167-186, 2007.
11. IS 1893:2016, Criteria for earthquake resistant design of structures.
12. IS 456:2000, Plain and reinforced concrete code of practice.
13. IS 875:2016, Code of practice for design loads.
14. IS 4991:1968, Criteria for blast resistant design of structures for explosions above ground.
15. Elements of Vibration Analysis by Leonard Meirovitch, McGraw-Hill Book Company.
16. Introduction to Structural Dynamics by J.M. Biggs, McGraw Hill Book Company.
17. Structural Dynamics, Theory and Computation by Mario Paz, CBS Publishers and Distributors.
18. Dynamics of Structures, Theory and Applications to Earthquake engineering by Anil K. Chopra, Pearson Education Asia.

Index

Absorber mass response, 119–120
Acceleration, 3–4, 6, 21, 79, 87–88, 140
    earthquake ground, 23
    spectral, 100
    time dependent, 85
ADRS plot, 99, 102–103
Amplitude
    of force, 26
    of oscillation, 9
Analysis model, 192–193, 195–196
ANSYS, 22
Arbitrary loading, 54
Asymmetric in plan, 197–205
Average spectrum, 91
Axial forces, 2
Axial hinges, 103
Axial load, 169

Bandwidth, 19, 110
Base isolators, 225
Base shear, 85, 90–91, 99, 102, 235, 274, 286
Beam retrofitting, 230
Bending moments, 2
Biaxial bending behavior, 166
Bilinear hysteric structure, 111
Blast resisting structure analysis and design
    characteristics of blast and consequences on structures, 277–278
    for explosions above ground, 279–281
    loading effects due to blasts, 278
    pressure-pulse shapes, 278
Bottom storey, 113–114, 121, 124–128, 130
Building asymmetric in plan, 197–205

Capacity spectrum, 98–105
Capacity Spectrum Manual from SAP2000, 102
CFST, *see* Concrete filled steel tube
Characteristic determinant, 46
Characteristic equation, 7, 46, 95
Circular frequency, 9
Collapse prevention (CP), 103
Column retrofitting, 230

Complex frequency response, 19
Complimentary function, 17
Concrete filled steel tube (CFST), 226, 233–234
Consistent mass matrix, 6
Constant acceleration, 80
Constant of proportionality, 16
Continuous systems, 4, 6
Coupling, 6, 61–62
    modal, 135–136
    torsional, 178
CP, *see* Collapse prevention
C++ programs, 236
C programs, 236
Critical damping, 81
Critically damped system, 12

D'Alembert principle, 4–6
Damped free vibrations, 11–16
Damped natural frequency, 13
Damped oscillation period, 16
Damped SDF system, 11, 26
Damping
    factor, 26
    matrices, 93
    ratio, 40–41, 48, 52
Deep earthquakes, 83
Degrees of freedom (DOF), 4, 282
    multi-degree of freedom system, 6
    single, 5–6
Design spectrum, 91–92
Differential equations, 2
    of equilibrium, 6
    of motion, 6, 34, 96, 98
Discrete systems, 4
Displacements, 2–3, 88
    vector, 93
Distant blast, 1
DLF, *see* Dynamic load factor
DOF, *see* Degrees of freedom
Drag effects, 277–278
Drag forces, 278
Drag pressure, 278
Ductility factor, 114

305

Duhamel's Integral, 53–59
Dummy time variable, 53
Dynamic
    analysis, 2–3, 43, 84
    loading, 1
    loads, 1, 55
    magnification factor, 20
    pressure, 278
    system modelling, 3–4
Dynamic load factor (DLF), 19, 69

Earthquake, 21
    force, 99
    ground acceleration, 23
    loading, 56
Earthquake ground motion, 1, 83
    response of RCC asymmetric buildings, 281–295
Earthquake resistant design, 83
    analysis, 209–217
    analysis model, 192–193, 195–196
    building asymmetric in plan, 197–205
    capacity spectrum, 98–105
    floating columns, 191
    hotel building analysis model, 189–191
    hotel building structural plan, 190
    mass participation factor, 206–208
    problems, 217–224
    response spectrum, 85–98
    shear building, 84–85
    soft storey, 197
    structural analysis, 83
    structural model, 84
    structural plan, 192–195
    structural system, 209
Earthquake response
    of elastic structure, 134–136
    of hybrid building, 275–277
    of space structures, 165
Eccentricity, 133, 142
    envelopes influence of maximum frame ductility, 144–146
    hysteretic energy *vs.*, 156
    influence, 174–180
Eigenvalues, 7, 93
    problem, 94
Eigenvectors, 46, 73, 76, 93
Elasto-plastic single-storey, 166

El Centro 1940 ground motion, 85–91, 144, 166, 285
El Centro earthquake, *see* El Centro 1940 ground motion
Energy dissipation due to hysteresis, 156–161
EOM, *see* Equations of motion
Equations of motion (EOM), 4, 7, 11, 24, 43, 49, 61, 112–113, 170–172
    inelastic torsional response, 139–144
    in non-dimensional form, 112
    parameters, 173–174
    solutions, 172–173
ETABS, 22, 189, 197, 206, 229, 273–275, 279, 283–284
Expendable top storey, 107–108, 111
Explosion, 1
Exponential decay, 19
External dynamic forces, 61–65

FBD, *see* Free body diagrams
Finite element method (FEM), 84, 237
Floating columns, 191
Floor stiffnesses, 92
Force, 49
    in absorber spring, 110
    exercise problems, 51–52
    transmitted to foundation, 50
    transmitted to support, 49–52
    vector, 93
Force-displacement relationship, 113, 168
Forced response of damped two-storey building, 46–48
Forced vibration, 7
    of damped SDF, 17–20
    SDF system, 17
Forcing frequency, 49
Fortran IV, 172
Free body diagrams (FBD), 11, 35
Free body forces, 36
Free vibrations, 7–10
    analysis of existing building, 234–236
    problem, 45
    problems in Python, 244–245
Frequency domain, response in, 25–28
Frequency response, 27
    spectrum, 110
Fundamental frequency, 151, 154
Fundamental mode, 151, 154–155, 164
Fundamental natural frequency, 45

Index

Gap element, 284
Generalised coordinate, 4
G+9 RC building retrofitting, 226
Ground acceleration, 67, 81, 85–88, 285
Ground floor, 108, 191, 209, 286
Ground motions, 65–68, 143
GT STRUDL, 206

Half-power bandwidth, 19
Harmonic forced oscillation, 51
Harmonic functions, *see* Sinusoidal functions
Harmonic loadings, 53
Harmonic motion, 30, 43, 45, 93
Hertz (Hz), 9
Higher modes, 92, 198
Hotel building
    analysis model, 189–191
    structural plan, 190
Hybrid building under seismic forces, 274
    connections, 274
    earthquake responses of, 275–277
Hysteresis, 156–161
Hysteric energy dissipation, 124–130
Hysteric vibration absorber, 111

IBM 360/44 computer, 172
Immediate occupancy (IO), 103
Impedance function, 27
Impulse loading, 54
Incremental velocity, 54
Inelastic earthquake response analysis, 166
Inelastic interaction effects, 165
Inelastic response spectrum, 90–91
Inelastic torsional response; *see also* Three degree of freedom system; Two degree of freedom system (2DOF system)
    column period, 180–181
    computer program, 144
    earthquake response of elastic structure, 134–136
    eccentricity envelopes influence of maximum frame ductility, 144–146
    eccentricity influence, 174–180
    energy dissipation due to hysteresis, 156–161
    equations of motion, 139–144, 170–174
    influence of strengthening exterior frames, 149–151
    maximum ductilities, 172, 175–180
    maximum ductility response spectra, 161–163
    P-Δ effect, 144, 149
    response history curves, 151–156
    structural model, 136–139, 166–167
    time-response curves, 181–183
    two-dimensional motion of mass center, 184–185
    two-dimensional motion of roof slab, 186
    yielding behavior, 167–170
    yield strength and period influence, 180–181
    yield strength influence, 146–149
Inelastic vibration absorber, 107–108
    absorber mass response, 119–120
    analysis of results, 114–119
    cumulative hysteretic energy, 128–129, 131
    displacement response-time curves, 121–124
    frequency ratio *vs.* maximum ductility, 119–120
    hysteric energy dissipation, 124–130
    maximum ductility contours, 115–118
    maximum ductility response spectra, 130–132
    numerical studies, 113–114
    ratios of maximum ductility and hysteric energy dissipation, 125–127
    response history curves, 120–124
    viscous damping influence, 130
Inertia force, 4–5, 22, 87, 134
Initial condition, 9–10, 14–15, 17–19, 48, 54–55, 59, 80
Inverse transformation, 30
IO, *see* Immediate occupancy
IPython, 240
Irregularity, 295
    stiffness, 284, 292, 295
    torsional, 207, 224, 290, 292–293, 295
IS 1893, 196
IS 1893:2016, 133, 189, 197
IS 4991:1968, 277–281

Jacketing, 225–236
Java, 236

Kinetic energy (KE), 24

Laplace transformation, 26
    method for SDF systems, 28–30
    response in, 25–28
Laplace variable, 30
Lateral force, 22, 87, 97, 99, 134, 189, 197
Life safety (LS), 103
Linear elastic period, 143
Linear elastic vibration absorber, 108–111
Load
    axial, 169
    dynamic, 1, 55
Loading effects due to blasts, 278
Logarithmic decrement, 16
LS, *see* Life safety
Lumped-mass model, 282

Machine learning (ML), 237
Magnification factor (MF), 19, 28
Mass matrix, 74, 93
Mass participation factor, 206–208
matplotlib, 239
Maximum deflection ductilities, 114
Maximum ductility response spectra, 130–132, 161–163
Maximum modal response, 80
Maximum pseudo-velocity, 85
Member-level retrofit, 225
MF, *see* Magnification factor
MIDAS, 206
ML, *see* Machine learning
Modal analysis
    multi-degree of freedom systems, 61–65
    multi-storey building, 65–68
    solved problems, 68–80
Modal coupling, 135–136
Modal eigenvectors, 64
Modal equations, 80
Modal force, 63
Modal mass
    of mode, 92
    participating factors, 286–288

Modal maxima, 64–65, 78
Modal stiffness, 63
Modal time period, 285
Modal vectors, 73
Mode participation factor, 67, 93
Modes, 45, 64, 92, 154–155, 198, 206
Mode shapes, 93
Mode superposition method, 61, 85, 94
Multi-degree of freedom system, 6, 61–65
Multi-storey building, 65–68

Natural frequency, 9, 17, 36–38, 46, 68, 93, 234
    Rayleigh's method to obtaining, 23–25
Natural period, 1
Newmark-Beta methods, 85
Newton's Law of Motion, 35–36, 53
Non-uniformity in strength of frames, 144
Nonlinear elastic, 2
Normal mode, 121, 127, 134–135
Number of zero crossings, 1
NumPy, 239

One-storey building, 21
Ordinary Moment Resisting Frame (OMRF), 283
Orthogonality, 92
    condition, 64
    functions, 67
    principle, 62
    with respect to mass, 64, 92, 94, 96
    with respect to stiffness, 94, 96
Over damped system, 13
Overpressure, 58

Pandas, 240
Participating factor, 92–93
Participation factors, 93
Particular integral, 17, 55, 57
P-Δ effect, 144, 166
    influence, 149
PE, *see* Potential energy
Peak ground acceleration, 85–87, 285
Period influence, 180–181
Phase angle, 9
Potential energy (PE), 24
Principle of conservation of energy, 24
Pseudo-acceleration, 86, 88–90
Pseudo-velocity, 88–90

Index

Pushover, 99–100
Pushover curve, 99–100
Pygame, 239
Python, 225, 236
    application in civil engineering, 237
    architecture, 237–239
    free vibration problems, 244–245
    history, 236
    libraries, 239–240
    response spectrum analysis of building using, 245–274
    static loading problem using, 240–244
pywin32, 239

Quality factor, 19

Ramp functions, 53
Random base motion, 1
Random harmonic loading, 53
Random inputs, 110
Random loading, 1
Random sinusoidal function, 6
Random sinusoidal motion, 23
Rate of loading, 1
Ratio of amplitudes, 16
Rayleigh's method to obtaining natural frequency, 23–25
Reinforced concrete buildings (RCC buildings), 225, 231
    Python, 225, 236–274
    response of RCC asymmetric buildings, 281–295
    retrofitting methods in reinforced concrete structures, 225–236
Resonance, 19–20
Response history curves, 120–124, 151
Response of RCC asymmetric buildings, 281
    modelling and analysis of structural irregularities, 283–295
    structural modeling, 282–283
Response spectrum (R–S), 85–98
    analysis using Python, 236–274
    method, 83, 87
Retrofitting methods in reinforced concrete structures, 225
    addition of shear walls, 225
    addition of steel bracing, 225
    base isolators, 225
    member-level retrofit, 225

    section enlargement or jacketing, 225–236
    structure-level retrofit, 225
ROBOT, 206
Root Mean Square value (RMS value), 65
R–S, *see* Response spectrum

Safe bearing capacity (SBC), 226–227
SAFE software, 279
SAP, 206
SBC, *see* Safe bearing capacity
SciPy, 239
SDF systems, *see* Single degree of freedom systems
SDOF, *see* Single degree of freedom systems (SDF systems)
Section enlargement, 225–236
Seismic resistant design, *see* Earthquake resistant design
Shear building, 84–85
Shear forces, 2
Shear walls, 133, 135, 197, 209, 218–219, 223, 226
Simple Harmonic Motion, 93–94
Single degree of freedom systems (SDF systems), 5–7, 22, 35, 92, 282
    damped free vibrations, 11–16
    exercise problems, 40–42
    at extreme positions, 25
    forced vibration of damped SDF, 17–20
    free vibrations, 7–10
    Laplace transformation method for SDF systems, 28–30
    logarithmic decrement, 16
    modelling procedure, 8
    Rayleigh's method to obtaining natural frequency, 23–25
    response in frequency domain and Laplace transformation, 25–28
    solved problems, 30–40
    subjected to support motion
    subjected to support motion, 21–23, 89
Single storey structure, 120
Sinusoidal functions, 6
Soft storey, 197
    failure, 292
Spectral acceleration, 100

Spectral velocity, *see* Pseudo-velocity
Speed, 1
Spring Mass system, 41, 66
Springs, 5, 112
Square Root of Sum of Squares (SRSS), 65, 80
STAAD PRO, 22, 274
Static analysis, 2–3, 84–85
Static deflection, 18
Static displacement, 73, 78
Static loading problem using Python, 240–244
Steady-state amplitude, 39
Step functions, 6
Stiffness
    irregularity, 284, 292, 294–295
    matrices, 93
    matrix, 74
Storey drift, 235
Strains, 2
Stresses, 2
Structural analysis, 83
Structural dynamics, 2
Structural irregularities, modelling and analysis of, 283–295
Structural model, 84, 112–113, 136–139, 166–167, 282–283
Structural plan, 192–195
Structure-level retrofit, 225
STTAD PRO, 206
Superposition
    method, 94
    of responses, 165
Support displacement, 49
Support motion, 21–23, 89
SymPy, 239

Three degree of freedom system; *see also* Inelastic torsional response
Three-dimensional model (3D model), 282
Three-storey
    building, 96
    structures, 166
Time-response curves, 181–183
Time-response relationships, 182–183
Time domain response, 25

Time function, 26
Torsion, 133–134, 163, 176, 197, 245, 282
Torsional coupling, 178
Torsional irregularity, 207, 224, 290, 292–293, 295
Torsional motions, 134–136
Torsional response, 134
Torsional rotation, 177
Total energy, 24
Transfer function, 29
Transient peak, 128
Translational motions, 134–136
Transmissibility, 50
    frequency ratio, *vs.*, 51
Triangular Pulse (Blast Load), 58
Tripartite plot, 85–86
Two degree of freedom system (2DOF system), 43–46; *see also* Inelastic torsional response
    exercise problems, 48
    forced response of damped two-storey building, 46–48
    two-storey frame, 44
Two-dimensional model (2D model), 282
Two-storey building, 112

Uncoupled equations, 72, 76
Undamped SDF system, 57

Vehicles travelling, 1
Velocity, 3, 88
    vector, 93
Vibration
    absorber principle, 110
    isolation, 49–50
Viscous damping, 16
    influence, 130

Wilson-Theta methods, 85

Yield
    acceleration of frames, 143
    strength influence, 146–149, 180–181
    surface, 169
Yielding behavior, 167–170